COFFEE, COCOA AND TEA

K. Willson
School of Biological Sciences
University of Liverpool
Liverpool, UK

CABI *Publishing*

CABI *Publishing* – a division of CAB INTERNATIONAL

CABI *Publishing*
CAB INTERNATIONAL
Wallingford
Oxon OX10 8DE
UK

Tel: +44 (0)1491 832111
Fax: +44 (0) 1491 833508
Email: cabi@cabi.org

CABI *Publishing*
10 E 40th Street
Suite 3203
New York, NY 10016
USA

Tel: +1 212 481 7018
Fax: +1 212 686 7993
Email: cabi-nao@cabi.org

A catalogue record for this book is available from the British Library, London, UK.

Library of Congress Cataloging-in-Publication Data

Willson, K.C. (Ken C.)
Coffee, cocoa and tea / K. Willson.
p. cm. – – (Crop production science in horticulture series ; 8)
Includes bibliographical references and index
ISBN 0-85198-919-5 (alk. paper)
1. Coffee. 2. Cacao. 3. Tea. 4. Coffee Industry. 5. Cocoa trade. 6. Tea trade. I. Title. II. Series: Crop production science in horticulture : 8.
SB265.W55 1999 98-34250
633.7′.3– –dc21 CIP

ISBN 0 85198 919 5

Typeset in Britain by Solidus (Bristol) Ltd
Printed and bound in the UK by the University Press, Cambridge

Crop Production Science in Horticulture Series

Series Editors: Jeff Atherton, Senior Lecturer in Horticulture, University of Nottingham, and Alun Rees, Horticultural Consultant and Editor, *Journal of Horticultural Science*.

This series examines economically important horticultural crops selected from the major production systems in temperate, subtropical and tropical climatic areas. Systems represented range from open field and plantation sites to protected plastic and glass houses, growing rooms and laboratories. Emphasis is placed on the scientific principles underlying crop production practices rather than on providing empirical recipes for uncritical acceptance. Scientific understanding provides the key to both reasoned choice of practice and the solution of future problems.

Students and staff at universities and colleges throughout the world involved in courses in horticulture, as well as in agriculture, plant science, food science and applied biology at degree, diploma or certificate level will welcome this series as a succinct and readable source of information. The books will also be invaluable to progressive growers, advisers and end-product users requiring an authoritative, but brief, scientific introduction to particular crops or systems. Keen gardeners wishing to understand the scientific basis of recommended practices will also find the series very useful.

The authors are all internationally renowned experts with extensive experience of their subjects. Each volume follows a common format covering all aspects of production, from background physiology and breeding, to propagation and planting, through husbandry and crop protection, to harvesting, handling and storage. Selective references are included to direct the reader to further information on specific topics.

Titles Available:

1. **Ornamental Bulbs, Corms and Tubers** A.R. Rees
2. **Citrus** F.S. Davies and L.G. Albrigo
3. **Onions and Other Vegetable Alliums** J.L. Brewster
4. **Ornamental Bedding Plants** A.M. Armitage
5. **Bananas and Plantains** J.C. Robinson
6. **Cucurbits** R.W. Robinson and D.S. Decker-Walters
7. **Tropical Fruits** H. Nakasone and R.E. Paull
8. **Coffee, Cocoa and Tea** K. Willson
9. **Lettuce, Endive and Chicory** E.J. Ryder

Contents

Section II: Coffee

Section III: Cocoa

Section V: Crop Processing

Preface

Although the amounts of the three 'stimulant' crops, coffee, cocoa and tea, produced and traded in the world are much smaller than the quantities of food crops, they are very important and an integral part of the culture of many countries. By their nature they have to be produced in tropical or subtropical areas. Their markets were developed, initially, mainly in temperate countries so major trading patterns developed in balance with their production by plantation systems. The involvement of organizations in 'developed' countries led to major research efforts in many countries. Relative to the quantities produced, the amount of research carried out on these crops is probably significantly larger than on many important crops produced in greater amounts. There is, therefore, a substantial volume of relevant research publications.

The processing of these crops after harvest is very important; users expect that the delicate flavours are retained until the beverage is drunk. The agronomic systems employed must link with the processing of the crop so that the flavour is retained in the final product. Interest in the final product has increased with concern about the effects of the stimulant compounds, particularly caffeine, on the human body. Much research has been carried out in this area. The general conclusion is favourable; most normal people can take in reasonable quantities of any one of these crops without damage in the long term. In the field these crops are environmentally friendly; permanent tree cover helps to prevent soil loss and degradation. Such degradation would result in loss of crop yield. Plantation systems minimize the loss of rainwater as runoff, and maintain the evaporative loss of moisture at a level similar to that of the forest which they have replaced, thus helping to avoid the desertification which is reducing the land area available for crop production. Major faunae are unwelcome, although at some places they sometimes stray into plantations. Minor faunae often continue to live within plantations.

Each crop produces a different part for harvest. In consequence, although the basic systems of growth, nutrition and propagation are the same, the practical methods from propagation to harvest vary substantially. Each crop has therefore been described in a separate group of chapters in this volume. Many major aspects vary widely; the coffee is harvested as small berries, cocoa produces large pods from which the beans have to be removed, while the crop from tea is leaves.

In many areas, cocoa and tea crop continuously, whereas coffee has in most areas one, but in some areas two, relatively short seasons for harvesting. Such effects have major implications for the management of plantations.

While there are many important pests and diseases that can affect these crops worldwide, each one has its own unique collection of crops and diseases, many of which are restricted to one or two continents only. The degrees of loss of crop due to pests and diseases vary; cocoa suffers more than any other crop in the world, losing an estimated 44% of potential crop. Coffee loses less, and tea the least.

Finally, and a matter of considerable importance, plantations of these crops provide stable employment for many indigenous people. Stable communities settle around plantations. The products are, in the main, exported, producing a vital income of foreign currency for the country concerned.

Stimulant Beverage Crops

NON-ALCOHOLIC STIMULANTS

A number of crops are classified as 'stimulants'. They all form substances which cause mild stimulating reactions in humans. They are neither 'drugs' in the sense that we use the word for plants such as cannabis and the opium poppy, which can be fatal when taken in excess, nor does the stimulant effect of these crops rely on alcohol.

The most important of the stimulant crops, apart from coffee, cocoa and tea, are the areca or betel palm *Areca catechu* which is distributed worldwide, and the cola nuts of West Africa, which are several species of *Cola*. Neither betel nor cola enters significantly into international trade, whereas coffee, cocoa and tea are traded in large quantities. In South America, the leaves from the mate plant (*Ilex paraguariensis*) are infused to produce a drink like tea. *Ilex* leaves contain from 0.9 to 2.2% caffeine; the amount diminishes as the leaves age (IARC–WHO, 1991a).

STIMULATING INGREDIENTS

The stimulating properties of tea, coffee and cocoa arise from the three compounds caffeine, theobromine and theophylline, which are all derivatives of purine. Table 1.1 lists the amounts of each of these substances in the normal products from the three crops.

Caffeine is considered to be the least desirable of these three compounds and, as a consequence, decaffeination of coffee and tea is now carried out on a large scale. The caffeine so produced is valuable as a drug for medicinal purposes and is added to stimulant drinks normally drunk cold, such as the various 'cola' formulations.

It has recently been reported (Coghlan, 1998) that the world's first caffeine-free coffee plants will soon be planted. Scientists at the University of

Table 1.1. Approximate proportions of purine derivatives (%) in coffee, cocoa and tea.

	Tea (dried leaves)	Arabica coffee (dried beans)	Robusta coffee (dried beans)	Cocoa (dried beans)
Caffeine	3–4	1.5	2.5	1.5
Theobromine	0.15	Trace	Trace	1.8
Theophylline	0.05	Trace	Trace	Trace

Hawaii have identified the master gene that governs caffeine production and have succeeded in blocking its function using an antisense gene carried by a bacterium. Experimental plants of *Coffea arabica* produced just 3% of the normal amount of caffeine. If all goes to plan, it will be possible to grow plants which will not need to be decaffeinated chemically in processes that can damage beans and spoil their flavour (see Chapter 4). In theory the same techniques could be used to develop caffeine-free tea, as the same gene controls caffeine production in tea.

HUMAN CONSUMPTION OF STIMULANT COMPOUNDS

Tea has the highest caffeine content, while that of cocoa is very small. Robusta coffee contains significantly more than arabica coffee. Although tea contains so much caffeine, the quantity of caffeine in a normal cup of tea is less than that in a cup of arabica coffee. This is due, at least in part, to the fact that less weight of tea is used per cup than coffee. The World Health Organization quotes a range of figures for the caffeine content of various beverages, as summarized in Table 1.2 (IARC–WHO, 1991b).

Estimates of per capita consumption of caffeine are given in Table 1.3. The amounts recorded as consumed are not necessarily totally ingested.

FLAVOUR PRECURSORS

A large number of chemical substances are present in each crop. During processing, many of these take part in reactions which produce the characteristic flavour of each. The chemistry of coffee flavour and aroma is particularly complex; apart from the purine derivatives, chlorogenic acids are important, as are polyphenols, which are oxidized by polyphenol oxidase. Reactions involving polyphenols are important also in cocoa fermentation, as is the conversion of carbohydrates to various acids. Polyphenols ('flavanols'), derivatives of catechin, contribute directly to the

Table 1.2. Ranges of reported values for caffeine in one cup (5 fl. oz, 148 ml) of various beverages (mg) (from IARC–WHO, 1991b).

Tea	Coffee	Cocoa	Soft drinks
27–60	59–115	4–42	0–29

Table 1.3. Per capita consumption of caffeine in USA and Great Britain, in grams per annum, for year 1982 (from IARC–WHO, 1991c).

Country	Tea	Coffee	Cocoa	Soft drinks	Others
UK	118	32	*	*	15
USA	13	46	2	13	5

*Included in 'Others'.

flavour of green (unfermented) tea. They are changed during fermentation to theaflavins and thearubigins, which give colour and flavour to black tea. The number of volatile substances in each crop is in the hundreds; the complex mixture thereof determines the aroma. The approximate composition of categories of chemical compounds in tea has been reported by E.A.H. Roberts as quoted by Robertson (1991).

EFFECTS ON HUMANS

While coffee and tea are used almost entirely as beverages, a large proportion of the cocoa produced is consumed as solid chocolate, which has a high nutritional and energy content. The physiological effects of these beverages on humans have been studied widely. Debry (1994) reviewed in detail many reports on the medical effects of coffee drinking. His lengthy conclusions may be summarized by reporting that normal healthy people may drink reasonable quantities of coffee without significant risk to their health.

However, coffee does contain substances that cause the level of cholesterol in the human body to rise. Stensvold *et al.* (1996) ran a long-term study of the cholesterol levels and deaths from heart disease of persons drinking a range of quantities of coffee. The most recent data showed a fall in the proportion of heavy coffee drinkers dying from heart disease. They attributed this to an increased proportion of coffee beverage being prepared by a filtration method rather than boiling without filtration. Urgert *et al.* (1996) compared the concentrations of deleterious substances in coffee beverage prepared with and without filtration. Filtration produced a significant reduction in the quantities of these substances in the beverage. Robert (1990) concluded that cocoa and chocolate are much less dangerous than tobacco, alcohol, tranquillizers and anxiolytics. Marks (1991) stated that 'the role of tea in causing any serious medical condition has not been established'. Current tea publicity emphasizes that tea is 'green'. It is a pure natural product with no additives and is always relatively fresh, as it is never stored for a long time. The quantity of chemicals used on tea in the field is low compared with many crops. Very recent work has shown that chemicals present in tea have significant effects as anti-ulcer, anti-diabetic, anti-cholesterol and anti-cancer agents (S. Sarma, 1996, brief summary of papers presented to the '95 International Tea Quality–Human Health Symposium, Shanghai, China, November 1995, personal communication). It has also recently been reported that, among men in the age range 50–69, the incidence of strokes was significantly lower, by 69%, in those who drank an average of 4.7 cups of tea per day than in those who drank 2.6 cups per day. This effect is caused by the flavonoids in tea (Anon., 1996). Landau and Yang (1997) have reviewed the effect of tea on health, showing that recent research has demonstrated beneficial results with laboratory animals. The extension of this work to humans is in progress; the results with animals suggest that few unfavourable and several beneficial effects can be defined. The German Medical Information Services (1997) have published the papers given to a Conference on the Chemical and Biological Properties of Tea Infusions. While few specific beneficial results are reported, there are indications that future research may well quantify such relationships.

Therefore there is no reason to fear that production, trade and consumption of coffee, cocoa and tea will not continue indefinitely to the benefit of everyone from producer to consumer.

ORIGINS AND HISTORY

All three crops have a long history of usage in the lands of their origin. These histories have given rise to numerous legends and some traditional customs. Various properties are also traditionally ascribed to these plants, some of which are rather apocryphal.

Coffee

Although arabica coffee originated, and can still be found, in the forests of Ethiopia, it was first exploited by the Arabs. It was, and still is, grown in the Yemen. Europeans first saw coffee in Arabia – hence the name 'arabica'. The earliest written reference to coffee is in 10th-century Arab writings. The oldest and best-known legend is the one about Mohammed, who was given coffee by the angel Gabriel, which gave him enough strength to unseat 40 horsemen and make love to 40 women. The Arabs guarded their coffee and ran a lucrative trade with Europe for 100 years from 1615. Eventually viable beans and plants were smuggled from Arabia. Coffee has since been planted throughout the world; its introduction to Great Britain as a drink led to the proliferation of 'coffee-houses' in London. These became meeting places for people from many walks of life and some developed into respected business institutions, such as Lloyds of London. As a beverage in general use, it was overtaken by tea in Britain. Coffee became the major beverage for the USA following the Boston Tea Party, which was a protest against the taxing of tea entering the then British colony. European mainland nations also consume large quantities, some at a higher rate per person than the USA.

Cocoa

Cocoa was revered by the indigenous inhabitants in the region of its origin, Central and northern South America. The myth was that the god of air, Quetzalcoatl, left the 'quachahuatl' tree to humans after he had been driven out of the Garden of Eden. Beans were used in religious ceremonies and as currency. A drink, 'xocoatl', was made by roasting beans and grinding them coarsely. Maize and spices were added and honey could be added as a sweetener; the mix could be fermented to make an alcoholic drink. As with coffee, aphrodisiac properties were attributed to cocoa; Montezuma was given a cup before entering his harem. Cocoa was imported into Spain and kept secret there for almost a century before spreading into the rest of Europe. The original formulation was not very popular in Europe; the taste was improved by adding sugar and omitting the spices and maize and its popularity, as the drink called 'chocolate', gradually grew; it became available in the British coffee-houses. Factory preparation of drinking chocolate started in the second half of the 18th century. Early in the 19th century, it was found that some of the fatty cocoa butter could be pressed out, which improved drinking chocolate. The cocoa butter had no value initially but was later found to be valuable in producing eating chocolate, which consists essentially of ground roasted beans with extra cocoa butter and sugar. A later development was to add milk to the mixture to produce milk chocolate.

Tea

The earliest authenticated use of tea was in southern China, the centre of origin of *Camellia sinensis* var. *sinensis*. Initially its uses were medicinal, but it was later found that an enjoyable beverage could be made from the unfermented leaves (green tea). The drinking of tea became a ceremonial ritual, particularly in Japan, where the custom is still important. Fermented (black) tea developed later and now forms a very high proportion of the tea consumed in Western countries.

Environmentally, well-managed tea plantations retain many of the characteristics of the tropical rain forest, which it replaces, particularly water absorption and retention and evapotranspiration (Willson, 1976). Plantations of cocoa and of coffee under most systems of management are similarly beneficial.

2

Origins and Distribution of Coffee, Cocoa and Tea

The species which give these three tropical crops developed in three different continents: coffee in Africa, cocoa in South and Central America and tea in Asia.

COFFEE

The exact number of species within the genus *Coffea*, which is in the family *Rubiaceae*, is not known, but is probably around 90. They are found in the centre of Africa, from the Congo basin to the highlands of Ethiopia. Two species have become significant in world trade: *Coffea arabica* and *Coffea canephora*. *Coffea liberica* is produced and traded on a very small scale.

C. arabica developed in the highlands in the south-east of Ethiopia (Ferwerda, 1976). In this region there are a number of variants, some of which have been considered to be subspecies. The genetic resources of Ethiopia have been reviewed by Mesfin Ameha (1991). Two important varieties are *C. arabica* var. *arabica* (syn. var. *typica*) and *C. arabica* var. *bourbon*. Most of the commercial coffee plantings in the world are of var. *arabica* but an increasing proportion of new plantings use mutants which have arisen and new varieties which have been developed from these. Various dwarf varieties are particularly important in this respect.

C. canephora is found throughout the rain forest of the Congo River basin and on the higher land to the north-east as far as the Ugandan shore of Lake Victoria up to about 1500 m altitude. It also occurs in the coastal rain forest from the Congo area as far west as Côte d'Ivoire. Within this vast area several distinct types are known but all are considered to be within the species *C. canephora*. The coffee produced by all the various types is known as robusta coffee. Two of the many types are worthy of mention. The upright type develops into trees; they have larger leaves and form bigger trees than *C. arabica*. This is considered to be the typical form so can be called var. *canephora*.

In contrast, the Nganda form has smaller leaves and more flexible stems, which bend outwards so that the plant develops sideways as well as upwards to produce a dome-shaped shrub.

C. liberica originated on the west coast of West Africa, in and around Liberia. Seeds have been spread by local people so it is now found throughout West Africa. Small plantings were made in Sri Lanka, Surinam, Indonesia and Malaysia. A very small amount of coffee from this species enters world trade; it is used locally and exported to countries where its bitter flavour is appreciated.

Coffea excelsa developed in drier conditions on the northern fringe of the West African rain forest. It produces poorer coffee than *C. liberica* but is grown and used locally to a limited extent in West Africa where the environment is suitable.

C. arabica was probably used by the local inhabitants of the Ethiopian highlands but its commercial use started when var. *typica* was introduced into the Yemen. From Arabia it was taken to Java. From a single tree sent from Java to Amsterdam in 1616 progeny were distributed among a number of countries in the Caribbean and South and Central America. Progeny from these early introductions was taken to other countries in the region. This material was introduced to Jamaica in 1730. Grown at high altitude in the Blue Mountains, this became the cultivar Blue Mountain; when planted at lower altitudes in Jamaica, it is not called Blue Mountain. However, it was later introduced to Kenya and Papua New Guinea, where it is called Blue Mountain. From these first introductions has grown the industry which now produces approximately two-thirds of the world's coffee.

Brazil dominates the regional production and has a major influence on the world industry; production amounted to approximately 21% of the world output in 1992. Colombia, producing 14%, is the second most important producer. Most of this production is arabica but processing into saleable beans is mainly by the dry process in Brazil, whereas the wet pulping process is used in Colombia (see Chapter 19, p. 240).

Arabica coffee was introduced to India in 1840 and to Ceylon at about the same time. Established in the highlands in South India, a small industry continues. In Ceylon and Indonesia, the fungal disease coffee rust, *Hemileia vastatrix*, caused a great deal of damage. The coffee industry in Ceylon was wiped out and tea was planted in its place. In Indonesia, arabica was largely replaced by robusta coffee and a substantial industry continues. Var. *typica* from Amsterdam was sent to the Philippines, where a substantial industry now supplies a large domestic market and a diminishing export trade.

In Africa, the indigenous robusta was used by the local population in various ways. Arabica was introduced as var. *bourbon*, which went first to Réunion and thence to Kenya, Tanzania and Uganda. Var. *typica* from the Edinburgh Botanic Garden went to Malawi and Uganda. Seed from Aden, presumably var. *typica*, was later brought to Kenya, where it is known as cv. French Mission and is clearly different from the Bourbon. Highland areas in Kenya, Tanzania,

Uganda, Angola, Burundi, Cameroon, Ethiopia, Malawi, Rwanda, Zambia and Zimbabwe now produce arabica coffee. Robusta coffee is produced commercially at lower altitudes in Benin, Cameroon, Cape Verde, Central African Republic, Congo, Côte d'Ivoire, Equatorial Guinea, Gabon, Ghana, Guinea, Liberia, Nigeria, Sierra Leone, Togo, Uganda and Zaïre. Côte d'Ivoire is the largest African producer, with 4.1% of world production, all robusta or a hybrid of arabica and robusta.

In North America, Mexico is a large producer – 4.9% of world production. Var. *typica* was taken to Hawaii to start the small industry there.

HYBRIDS OF COFFEE

A few natural hybrids have been found but none has been used for commercial production. A hybrid between *C. canephora* var. *ugandae* and *Coffea congensis*, called Congusta, occurred spontaneously in Java and has properties which make it a potential replacement for robusta. A recent discovery was Hibrido de Timor, which is believed to be a natural hybrid between *C. arabica* and *C. canephora*. It is grown commercially on the island of Timor, where it was found, and is now used extensively in breeding programmes, as it is resistant to *H. vastatrix* and has a strong resistance to coffee berry disease, *Colletotrichum kahawae*.

Direct crossing of *C. arabica* and *C. canephora* produces triploid trees, which flower profusely but are sterile; fertile hexaploids can be obtained by doubling the chromosome number with colchicine. The chromosome number of *C. canephora* can be doubled by treatment with colchicine. This hybridizes with *C. arabica* to give fertile hybrids, which have been given the name Arabusta. These are being developed in the Côte d'Ivoire, with the aim of finding a hybrid that will give the quality of arabica coffee when grown in tropical lowlands.

COCOA

The genus *Theobroma*, in the family *Sterculiaceae*, includes over 20 species, but only *Theobroma cacao* has any commercial importance. Beans from *Theobroma bicolor* have been used to a small extent locally in South America, being probably harvested from wild trees. Cocoa from this species has a very low quality.

There are two types of *T. cacao*, sufficiently distinct to be regarded as subspecies. Criollo developed north of the Panama isthmus, while Forastero originated in the Amazon basin, which was probably the original home of the species: Criollo probably developed from Forastero that had moved northward (Cope, 1976). Criollo was used in South and Central America before the arrival of the Spanish, being cultivated in various places in Central America. The beans were used as currency and an unsweetened drink was made. This drink was not to European taste, but it was found that cocoa with sugar made a palatable drink. The first significant commercial production was in Brazil, where the Forastero type was brought into production. Most of the Forastero cocoa now planted is of a relatively uniform type called Amelonado.

The growing of cocoa spread to Caribbean islands, particularly Jamaica, Trinidad, Martinique and Haiti. The earliest plantings were probably of Criollo, but Forastero was introduced also. This led to hybridization; the distinct and well-known hybrid called Trinitario developed in Trinidad. The Spaniards took cocoa to the Philippines, whence it spread through South-east Asia and to India and Ceylon. Amelonado cocoa was later taken to West Africa. This developed into one of the major industries of West Africa.

The demand for cocoa was stimulated by the discovery, in 1825, that cocoa butter could be separated from the ground beans and subsequently mixed with more beans to produce solid chocolate. Hitherto, use had been limited to a beverage made from ground beans and sugar.

Cocoa is now grown in at least 16 countries in Africa, between Sierra Leone, Uganda and Angola, with production dominated by Côte d'Ivoire, Ghana, Nigeria and Cameroon. At least 19 countries in the Caribbean and South and Central America grow cocoa; Brazil is a major producer, while Ecuador is the next in importance; a distinctive type of cocoa called Nacional is grown in Ecuador. In the East, Fiji, India, Indonesia, Malaysia, Papua New Guinea, the Philippines, Samoa, Solomon Islands, Sri Lanka and Vanuatu produce cocoa.

TEA

The genus *Camellia*, in the family *Camelliaceae*, developed in Asia, centred around the Himalaya mountains. There is a wide spread of the various species; they have been valued for hundreds of years, either for their flowers or for medicinal and stimulant properties. In consequence, they have no doubt been carried out of their original habitats. Nevertheless, it is certain that the two major types of *Camellia* with stimulant properties developed separately: *Camellia sinensis* var. *sinensis* on the northern slopes of the Himalayas and *C. sinensis* var. *assamica* on the southern slopes and adjoining plains.

The China type, var. *sinensis*, developed at elevated altitudes in forest which was probably not as wet or dense as the southern forests which were home to the Assam type, var. *assamica*. This variety therefore developed as a bush rather than a tree, with small leaves and greater tolerance of drought and low temperatures. The various uses of this variety were developed by the Chinese and its cultivation spread over China and to Japan. The earliest exports of tea from China were overland to Russia. The western European countries first met tea when their early maritime explorations reached China. The export of tea to Europe became a major business.

The East India Company had, for many years, a monopoly of the China tea trade. Sir Joseph Banks suggested to the Company in 1788 that tea would grow on the southern slopes of the Himalayas, but there was no action until 1833, when the Company's monopoly of the tea trade ended. A committee was set up to investigate the viability of tea in Assam. One of its members, C.J. Gordon, was sent to China to acquire tea and tea-makers. In the meantime, Major R. Bruce had reported that plants similar to tea were growing wild throughout the forest in north-east Assam. These plants were not confirmed as *C. sinensis* until after Gordon had shipped seeds from China, from which plants were raised in the Calcutta Botanic Garden and then planted at several places in North India and in the hills of South India. The greatest success was in the Brahmaputra valley in Assam and this area was chosen for major development. Local plants were incorporated into the plantations. The two varieties, China and Assam, were therefore growing side by side from the beginning of the Indian tea industry (Weatherstone, 1986). More seeds were brought from China, but local seeds began to be used also. These must have been mostly, if not wholly, hybrids. In due course, orchards of trees, known as 'baries', were planted specifically to give seeds for propagation. Local seeds were used, which must have been hybridized to some degree. Tea which has been planted since, in most of the world's tea-growing regions, except China and Japan, originated from the Assam plantings and is therefore hybridized.

The Assam plantings grew to become the largest area of tea in the world. Of the experimental plantings in the hills, the area around Darjeeling grew to become important although small in area. In South India, significant development of tea did not begin until almost the end of the 19th century. Tea is also grown on a large scale in Bangladesh.

The next major development was in Ceylon. The large coffee industry there was destroyed by leaf rust, *H. vastatrix*, in the second half of the 19th century and was replaced by tea. The Dutch planted a little tea in Java, using seed from China and Japan, early in the 19th century but there was no significant expansion until the end of the century, using hybrid seed from India and Ceylon. Indonesia is now a major producer of tea. Malaysia and Vietnam are also producers. Tea production in Africa started in the 1930s, primarily in East and Central Africa where the climate is very suitable, in the Equatorial highlands and at lower altitudes to the south. Burundi, Ethiopia, Kenya, Malawi, Mozambique, Rwanda, South Africa, Tanzania, Uganda, Zambia and Zimbabwe are all producers. In West Africa, Cameroon, Mali and Zaïre have planted tea, which is also grown on islands such as Madagascar, Mauritius, Réunion and the Seychelles.

In South America, production is reported from Argentina, Bolivia, Brazil, Ecuador and Peru.

Around the Black Sea, the tea industry in Georgia and some neighbouring countries has a long history. The earliest plantings, at least, were probably made using seed which came overland from China. It is a large industry, and its output was sold mainly within the former USSR. Turkey also has a large industry whose output is totally absorbed by the home market. Iran has a considerable area of tea. Australia has developed a very highly mechanized industry, while Papua New Guinea has several plantations in the highlands.

3

PRODUCTION AND MARKETS

PRODUCTION

Tables 3.1, 3.2 and 3.3 list the production data by country for coffee, cocoa and tea for the year 1996, as reported by the Food and Agriculture Organization (FAO). Coffee production is much higher than the production of the other two, which are at very similar levels. All three are major contributors to world trade. Although internal consumption of these crops is increasing in the producing countries, particularly of tea, the major proportion of these very large production totals enters into international trade.

Coffee

A large proportion of the countries which have a significant area of land with a suitable climate have planted coffee. Table 3.1 shows that 56 countries reported details of a significant level of coffee production. Production is dominated by Brazil, with 21.1% of the world's coffee in 1996. Together with Colombia, which produced 13.9% of the total, these two producers dominate the world market. Both produce primarily arabica coffee. Their nearest competitor, in terms of total production, is Indonesia, producing 7.3% of the world's total, but this is mostly robusta. The amount of coffee produced in Brazil, in particular, can have a significant effect on world price levels.

Cocoa

Production is, by weight, just under 50% of the production of coffee, but this is still a large quantity. Most of this total is traded around the world; local use in producing countries is relatively small. While the areas of origin, South and Central America, produce a substantial quantity, the production in Africa is

Table 3.1. Coffee: production data 1996 (from FAO, 1996a).

Country	Area harvested (1000 ha)	Average yield (kg ha^{-1})	Production (1000 t)
Angola	120*	42	5
Burundi	39*	726	25*
Cameroon	270*	195	53
Central African Republic	26*	500	10*
Congo	4*	368	1*
Côte d'Ivoire	1405*	118	165
Equatorial Guinea	19*	378	7*
Ethiopia	95*	780	230*
Ghana	10*	300	3*
Guinea	55*	410	23
Kenya	156*	626	98*
Liberia	20*	150	3*
Madagascar	193*	373	72
Malawi	4*	1275	5*
Mozambique	1*	500	4*
Nigeria	8*	357	3
Rwanda	40*	525	21*
Sierra Leone	14*	416	25*
Tanzania	125*	416	52
Togo	38*	421	16*
Uganda	280*	917	257*
Zaïre	240*	251	60*
Zambia	2*	884	2*
Zimbabwe	7*	1539	10*
Total: Africa	3171		1150
Costa Rica	101*	1412	143*
Cuba	85*	212	18*
Dominican Republic	165*	283	47
El Salvador	167*	771	126
Guatemala	269	500	207
Haiti	54*	630	27*
Honduras	209	630	131
Jamaica	6*	469	3*
Mexico	763	426	325
Nicaragua	84	656	55
Panama	32*	388	12*
Puerto Rico	30*	431	13*
Trinidad and Tobago	6*	139	2
USA	1	1022	2
Total: North and Central America	1972		1111

Table 3.1. Continued

Country	Area harvested (1000 ha)	Average yield (kg ha^{-1})	Production (1000 t)
Bolivia	24	995	23
Brazil	1984	650	1290
Colombia	965*	852	822
Ecuador	305	403	155
Paraguay	6*	671	4
Peru	180	592	107
Venezuela	354*	249	88*
Total: South America	3818		2489
China	22*	2046	45*
India	230*	783	180*
Indonesia	771	559	431
Laos	158	600	9*
Malaysia	15*	758	11*
Myanmar	4	397	1
Philippines	134	1110	149
Sri Lanka	17*	691	11*
Thailand	69*	1093	76*
Vietnam	152*	1303	198*
Yemen	29*	363	11
Total: Asia	1601		1122
Papua New Guinea	50*		60
Total: Oceania			60
Total: World	10,612		5932

*Unofficial figure.

the highest of any continent, amounting to 64.9% of world production. The country with the highest production is Côte d'Ivoire – 4% of world production. The crop has become popular in some Asian countries with high rainfall levels, notably Indonesia and Malaysia, and in a number of Pacific islands.

Tea

The weight of tea produced in the world is about the same as that of cocoa beans produced. While a good proportion of this is traded around the world, consumption in producing countries, particularly India, is increasing. Tea

Table 3.2. Cocoa beans: production data 1996 (from FAO, 1996b).

Country	Area harvested (1000 ha)	Yield (kg ha^{-1})	Production (1000 t)
Cameroon	360*	349	126
Congo	6*	246	2*
Côte d'Ivoire	2150*	584	1254
Equatorial Guinea	60*	75	5
Gabon	15*	81	2
Ghana	1200*	283	340
Guinea	5*	800	4*
Madagascar	6	729	4
Nigeria	400*	363	143*
São Tomé	24*	125	3
Sierra Leone	3*	3333	10*
Tanzania	4*	628	3
Togo	36*	111	4*
Uganda	11*	114	1*
Zaïre	22*	352	75
Total: Africa			1976
Costa Rica	20*	150	3
Cuba	9*	235	2*
Dominican Republic	137	235	63
Grenada	2*	1070	2*
Guatemala	5	645	3
Haiti	9	467	4
Honduras	3*	1000	3
Jamaica	4*	370	1
Mexico	91	578	53
Panama	4*	243	1*
Trinidad and Tobago	17*	113	2*
Total: Central America			137
Bolivia	6*	357	4*
Brazil	688	511	373
Colombia	124	522	65
Ecuador	350	250	88
Peru	36	624	22
Venezuela	65*	289	19*
Total: South America			571
India			7*
Indonesia	332	826	274
Malaysia	205	610	217
Philippines	7*	440	7*

Table 3.2. Continued

Country	Area harvested (1000 ha)	Yield (kg ha^{-1})	Production (1000 t)
Sri Lanka	6*	626	4*
Total: Asia			509
Papua New Guinea	85*	353	30
Samoa	3*		1
Solomon Islands			3*
Vanuatu	4*	530	2*
Total: Oceania			36
Total: World			3229

*Unofficial figure.

Table 3.3. Tea: production data 1996 (from FAO, 1996c).

Country	Area harvested (1000 ha)	Yield (kg ha^{-1})	Production (1000 t)
Burundi	6*	823	5*
Cameroon	2*	2042	4*
Ethiopia	3*	264	1*
Kenya	110*	2318	255*
Malawi	19*	1990	37
Mauritius	2*	1529	3*
Mozambique	4*	769	3*
Rwanda	7*	857	16*
South Africa	11*	1141	12*
Tanzania	19*	1211	23
Uganda	24*	708	17
Zaïre	9*	384	3*
Zambia	5	1250	1*
Zimbabwe	5*	3165	15
Total: Africa	220		395
Argentina	37	1075	40
Bolivia			3
Brazil	5*	2000	10*
Ecuador	1	2396	2
Peru	3	783	2
Total: South America	47		57

continued overleaf

Table 3.3. Continued

Country	Area harvested (1000 ha)	Yield (kg ha⁻¹)	Production (1000 t)
Azerbaijan	13*	308	4*
Bangladesh	48	997	48
China	872*	699	609*
Georgia*	46	742	34
India	428*	1671	715*
Indonesia	117	1442	169
Iran	32	1750	56
Japan	55*	1636	90*
Republic of Korea	1*	1400	1*
Laos	1*	2604	2*
Malaysia	3*	2000	6*
Myanmar	59	269	16
Nepal	1*	3525	3
Sri Lanka	197*	1247	246*
Thailand	17*	300	5*
Turkey	77*	1610	124
Vietnam	71*	568	40*
Total: Asia			2168
Europe	2	1515	1
Australia*			0.8
Papua New Guinea	5*	1957	9*
Oceania			9
Russian Federation	2*	1547	2*
Total: World	2310		2633

*Unofficial figure.
In addition to the countries listed above, tea-growing is being developed in Morocco (Banhammou, 1993) and New Zealand (Smale, 1991).

produced in the Middle East, Iran, Turkey and the former USSR is almost entirely consumed in the producing country, as is the small Australian output.

In 1992, 75.4% of world production was in Asia, the continent of origin of the species. Countrywise, the highest production was in India, with 27.3% of world production, closely followed by China, producing 20.9%. Sri Lanka, where tea is the country's most important crop, produced 9.4% of the world total. African tea production is increasing rapidly but it only amounted to 14.7% of world output. Small quantities of tea are produced in the former

USSR, South America, Papua New Guinea and Australia. Profiles of the tea industries in 13 major producers have been published by the Tea Research Association, India (1993).

Crop Yields

There is a very wide variation of crop yields, i.e. production per hectare, from one country to another in all three crops. For some smaller countries, the data for area planted and total yield are suspect; accurate data are not available from either government or trade sources. The organizations publishing the data give estimated figures, some of which may differ considerably from the exact values.

Apart from incorrect data, the figures indicate the average standard of husbandry in the particular country. High yields of all three crops can only be achieved by sufficient plants, evenly spaced, to give complete cover of the area planted, large inputs of essential nutrients and a high standard of husbandry, using the best practices in pruning, retention of vegetative litter in the plantation and effective pest, disease and weed control. Most commercial estates maintain a high standard to achieve high yields and reasonable profitability. Many smallholders plant a mixture of crops so as to produce food for subsistence, of which surpluses may be sold, and to obtain some cash income from tree crops. In many cases there will not be sufficient trees to give a complete cover of the tree crop, whence figures reported for planted areas will become inflated. Planting in partly cleared forest can result in uneven distribution. Given easy availability of technical advice and the necessary physical inputs, and a willingness to follow advice and submit to control of husbandry practices, smallholders can achieve high yields of a high-quality product. In Kenya, for example, such a policy has been very effective for tea.

Political factors, such as taxation policy and civil unrest, can have a major effect on production and yields.

MARKETS

Tables 3.4, 3.5 and 3.6 list the countries which imported the three crops in 1996 and the quantities of imports reported to the relevant international organization.

Coffee

Figures for members of the International Coffee Organization (ICO) (see p. 29) are quoted separately from those of non-members. Inevitably, data for

Table 3.4. Coffee: imports by receiving countries in 1996 (from International Coffee Organization, 1996a, b).

Country	Amount imported (1000 t)
Members of ICO	
USA	1376
Austria	132
Belgium/Luxemburg	109
Denmark	64
Finland	63
France	397
Germany	827
Greece	20
Ireland	6
Italy	276
Netherlands	192
Portugal	35
Spain	205
Sweden	106
UK	185
Total: EEC	2617
Cyprus	2
Fiji	0.06
Japan	32
Norway	44
Singapore	93
Switzerland	66
Total: other members	237.06
Total members of ICO	4230.06
Non-members of ICO (1990)	
Bulgaria	16
Czechoslovakia	40
Hungary	34
Poland	20
Romania	20
Former USSR	58
Yugoslavia	63
Others in Europe	6
Algeria	62
Morocco	21
Others in Africa	19
Israel	16
Democratic Republic of Korea	n/a (1 in 1989)
Republic of Korea	51

Table 3.4. Continued

Country	Amount imported (1000 t)
Lebanon	10
Saudi Arabia	14
Others in Asia	27
Canada	118
Others in North America	0.3
Australia	38
New Zealand	7
Others in Oceania	n/a
Other importing non-members	10
Producing non-members:	
Argentina	30
Malaysia	9
South Africa	17
Others	5
Total non-members	711.3
Total: World	4941.36

EEC, European Economic Community; n/a, not available.
Figures have been rounded up to the nearest digit, so the sum of the above figures does not agree with the actual grand total quoted.

non-members are published later than those for members and may be less accurate. All the major users are members so that the data show a good overall picture of the coffee industry.

A very high proportion of the world's coffee is imported by developed countries; consumption in the producing countries and other developing states is small. The largest individual importer is the USA, which imports 23% of the total world imports. The European Economic Community as a whole imports 39% of the world total; Germany imports most, over twice the amount of the nearest competitor, France, and approximately four and a half times the UK quantity.

Some coffee is traded between countries which have imported it. This is mostly between developed countries and does not alter the general conclusion that coffee is consumed mainly in developed countries. Some of the re-exports will have been processed, for example roasted, packaged or made into soluble ('instant') coffee. The value of coffee to developing countries is therefore as an export, which, in many cases, makes a significant contribution to the earnings of the foreign exchange which is necessary for the purchase of essentials.

Table 3.5. Cocoa beans: imports by receiving countries 1 April 1995 to 31 March 1996 (from International Cocoa Organization, 1997).

Country	Amount imported (t)
Austria	15,007
Belgium/Luxemburg	57,115
Denmark	3,703
Finland	120
France	117,127
Germany	299,124
Greece	1,500
Iceland	20
Ireland	9,190
Italy	71,207
Netherlands	404,643
Norway	51
Portugal	83
Spain	49,537
Sweden	8
Switzerland	19,763
UK	248,327
Yugoslavia	5,000E
Total: Western Europe	1,301,525
Bulgaria	4,000
Czech Republic	14,000
Hungary	5621
Poland	32,000
Romania	759
Former USSR	17,471
Slovak Republic	13,894
Total: Eastern Europe	87,745
Egypt	500E
Kenya	50
Morocco	90
South Africa	6,353
Tunisia	550E
Zimbabwe	0
Total: Africa	7,543
Argentina	3,000E
Canada	38,560
Chile	80E
Colombia	830E

Table 3.5. Continued

Country	Amount imported (t)
Costa Rica	910E
Guatemala	8
Nicaragua	40E
Panama	480E
Peru	12
United States	445,260
Uruguay	120E
Total: Americas	489,300
Australia	21E
China	33,000E
Hong Kong	50
India	1,660E
Iran	100E
Israel	1,800E
Japan	49,004
Republic of Korea	1,880
Malaysia	24,084
New Zealand	189
Philippines	9,890
Singapore	88,340E
Thailand	9,875
Turkey	17,942E
Total: Asia and Oceania	237,835

Figures have been rounded up.
E, estimated total.

Coffee contributes nothing directly to the supply of food for local consumption but is not incompatible with smallholder mixed farming.

Cocoa

The data in Table 3.5 show that approximately 44% of the world production of cocoa is imported by western Europe and 15% by the USA. Other relatively large importers are China, Japan, the Republic of Korea, Singapore and Canada. It is therefore the developed countries which consume cocoa in its various finished products. Considerable quantities are re-exported, as either beans, partly finished materials, for example cocoa butter, or finished products, such as chocolate. These re-exports do not significantly alter the main conclusion that the major proportion of the market for cocoa is in developed countries. Like

Table 3.6. Tea: imports for consumption by receiving countries 1996, adjusted for re-exports (from International Tea Committee, 1997a).

Country	Amount imported (t)
UK (net imports)	148,452
Austria	1,500
Belgium/Luxemburg	1,158
Denmark	1,757
Faroe Islands	100
Finland	1,200
France	14,973
Germany	24,284
Gibraltar	80
Greece	550
Iceland	60
Eire	11,896
Italy	5,000
Malta	550
Netherlands	7,907
Norway	1,104
Portugal	300
Spain	850
Sweden	3,000
Switzerland	1,277
Yugoslavia (Federal Republic – Serbia and Montenegro)	1,200
Total: Western Europe (excluding UK)	78,764
Bulgaria	700
Czech Republic	1,200
Slovakia	600
Hungary	1,597
Poland	26,500
Romania	200
CIS (former USSR)	44,000
Russian Federation	111,096
Baltic States	2,000
Total: Eastern Europe	187,893
Canada	13,492
USA	89,155
Bahamas	200
Barbados	120
Bermuda	110
Belize	36
Jamaica	198

Table 3.6. Continued

Country	Amount imported (t)
Trinidad and Tobago	170
Netherlands Antilles	60
Others in North America and West Indies	600
Total: North America and West Indies	104,141
Mexico	220
Central America	360
Argentina	90
Bolivia	100
Brazil	70
Chile	13,642
Peru	100
Uruguay	650
Other countries in Central America	800
Total: Latin America	16,032
Abu Dhabi	1,907
Bahrain	800
Dubai	17,983
Kuwait	4,600
Oman	2,300
Qatar	1,200
Saudi Arabia	14,100
Other Arab states	9,500
Afghanistan	48,000
Hong Kong	7,713
Iran	27,300
Iraq	2,000
Israel	3,000
Jordan	25,000
Lebanon	3,000
Malaysia	6,220
Nepal	800
Pakistan	110,703
Philippines	400
Sabah	300
Sarawak	400
Singapore	600
Syria	21,074
Thailand	600
Others in Asia	1,800
Total: Asia (excl. China, Japan and Taiwan)	311,300

continued overleaf

Table 3.6. Continued

Country	Amount imported (t)
Algeria	7,000
Benin	10
Botswana	1,500
Burkina Faso	170
Cameroon	20
Canary Islands	460
Central African Republic	4
Ceuta	55
Chad	900
Congo	30
Egypt	71,700
Ethiopia	1,800
Gabon	40
Gambia	1,400
Ghana	130
Guinea	150
Côte d'Ivoire	1,200
Lesotho	700
Libya	10,000
Mali	2,800
Mauritania	1,300
Mauritius	20
Melilla	100
Morocco	28,400
Niger	1,000
Nigeria	6,000
Senegal	5,500
Somalia	3,000
Republic of South Africa	12,136
Sudan	15,200
Swaziland	250
Togo	1,800
Tunisia	9,700
Zaïre	10
Zambia	160
Zimbabwe	2
Others in Africa	2,200
Total: Africa	186,847
Australia	17,630
Fiji	570
New Zealand	4,510

Table 3.6. Continued

Country	Amount imported (t)
Papua New Guinea	100
Others in Oceania	500
Total: Oceania	23,310
China	4,000
Japan	48,420
Taiwan	7,365
Total: China, Japan, Taiwan	59,785
Total: World imports for consumption	1,095,506

coffee, the value of cocoa to developing countries is as a source of foreign exchange. It is a crop that can be grown by smallholders and is not totally incompatible with subsistence farming.

Tea

From Tables 3.6 and 3.7 it can be seen that consumption of tea in the producing countries is greater than consumption in the countries which import tea for their own use. Consumption of tea in producing countries in 1991 was 57.4% of the sum of imports by consumers and tea retained by producers. This situation is markedly different from coffee and cocoa, where consumption of local production is a small proportion of total production.

World tea production has increased greatly since the Second World War. The rate of increase of production has been much greater than the rate of increase of consumption by importing countries. The difference has been absorbed by the producing countries for their own use. In spite of the rapid increase in production, world supply and demand have remained approximately in balance, thanks to the increasing consumption by producers.

The country which imported the most tea in 1991 was the USSR. The breakup of the USSR and the ensuing civil unrest in some parts has reduced this figure considerably. The imports were in addition to local production equal to about two-thirds of the level of imports.

The UK imported the next highest quantity of tea in the world: this quantity is outstanding among the Western nations, as might be expected. It was 1.6 times the amount imported by the USA and about six times the quantity imported by Germany, which was the continental European nation with the highest imports.

Table 3.7. Tea: quantities retained for local consumption by producing countries, 1996 (from International Tea Committee, 1997b).

Country	Amount retained (t)
India	626,350
Bangladesh	28,983
Sri Lanka	25,396
Indonesia	42,468
China	423,716
Taiwan	19,847
Iran	48,300
Japan	88,214
Malaysia	5,700
Turkey	112,540
Vietnam	25,000
Burundi	1,320
Kenya	12,936
Malawi	571
Mauritius	1,115
Mozambique	1,150
Rwanda	5,500
Republic of South Africa	9062
Tanzania	1,325
Uganda	2,436
Zaïre	500
Zimbabwe	5,252
Argentina	1,693
Brazil	3,000
Ecuador	700
Peru	2,200
Australia	1,142
Papua New Guinea	700
Total	1,497,116

Third highest in the world was Pakistan, which has not yet developed the production of tea, although there are districts where this would be possible. The popularity of tea has grown, as it has in neighbouring India; imports almost doubled between 1980 and 1991.

Arab countries import relatively large quantities. Afghanistan headed the list, with Egypt second. Iran, Jordan, Morocco, Libya, Syria, Saudi Arabia, Sudan and Dubai all imported over 10,000 metric tonnes (t) of tea in 1996.

Other countries which imported relatively large quantities are Japan, where imports increased by one-third from 1992 to 1996, and imports were just over one-half of local production. The latter was almost entirely consumed in Japan. Australia imported a substantial quantity, and Chile is a country with a small population which took in a relatively large amount.

Table 3.7 shows that in Turkey, where the industry exists primarily to meet local needs, 90.7% of production was consumed in the country. Next was India, retaining 87.6%, followed by China, retaining 69%.

MARKETING SYSTEMS

There are considerable differences between the ways in which the three crops are offered to the market and sold. For coffee, all major producers and most of the major importers are members of an international organization, which, in the past, set quotas for exports by each member country, so as to minimize price fluctuations. Coffee is sold by contracts between exporters and importers; two standard types of contract are used throughout the world.

Only in cocoa is there a substantial market in 'futures' as well as in 'actuals'; a number of standard forms of contract, drawn up by national or international bodies, are used for the various sales transactions.

At one time all tea was sold by auction. In recent years substantial quantities have been sold by contracts between exporters and importers. Nevertheless, all the auction markets still operate and they set price levels. The London auction market, although now handling only a small proportion of the tea produced in the world, is still accepted as the place where international price levels are set. It is the only auction where teas from different countries are offered for sale side by side so the sale prices set an international standard.

Coffee

International Coffee Organization

The ICO was formed in 1962, at a time when world output of coffee was increasing, largely due to countries which had hitherto produced little or no coffee planting coffee on a large scale. Most major producing and consuming (importing) countries are members. The ICO negotiated the first International Coffee Agreement (ICA), which limited production with the aim of preventing the price falling too low. Each member was given a 'quota' quantity to import or export. The ICO set an 'indicator price', which was a range within which all deliveries of coffee should be sold. The quota for each country was set with the aim of keeping the price within the price bracket.

Coffee could be sold to countries which are not members of the ICO at any price which could be negotiated. This was usually below the quota price, which led to a temptation to transfer such cheap coffee to member countries. Every shipment under quota had to be accompanied by a quota certificate that had been validated by carrying a special adhesive stamp issued by the ICO.

The ICA collapsed in 1989 under pressure of increased production and a demand by consuming countries for a free market in coffee. Negotiations to establish a modified agreement were set in motion but collapsed in 1993.

The price for coffee continued at a low level. In August 1993, 20 countries met and set up the Association of Coffee Producing Countries (ACPC). They agreed to withhold 20% of their production until prices reached an acceptable level, above which withheld stocks would be released (Anon., 1993). Countries present at the first meeting were Angola, Burundi, Brazil, Cameroon, Colombia, Congo, Côte d'Ivoire, Costa Rica, El Salvador, Ethiopia, Gabon, Guatemala, Kenya, Madagascar, Rwanda, Tanzania, Togo, Uganda, Zaïre and Zimbabwe.

The ACPC is setting up an office in Rio de Janeiro. In the meantime, the ICO office in London continues to collect and publish statistical data listing production, consumption and prices of coffee. The ICO maintains a library of relevant material and is involved in distributing technical and promotional material.

Cocoa

International Cocoa Organization

Most major producing, importing and processing nations belong to the International Cocoa Organization (ICCO). It has a head office in London, in the same building as the ICO. The ICCO was set up in 1973 to administer the International Cocoa Agreement (ICCA), which was negotiated in 1972 under the auspices of the United Nations Conference on Trade and Development (UNCTAD). The aim was to stabilize the price of cocoa, which had hitherto been very volatile.

The ICCO publishes the *Quarterly Bulletin of Cocoa Statistics*, which lists not only production, stocks, exports and imports of cocoa beans but also the quantities of cocoa processed in various countries and the amounts of the intermediate and final products: cocoa butter, cocoa powder and cake, cocoa paste/liquor, chocolate and chocolate products, imported and exported. The head office has a library of relevant material and handles technical and promotional literature.

International Cocoa Agreement

The 1972 agreement fixed a range of prices within which cocoa should be sold. The mechanisms to achieve this were export quotas for producers and purchase of a buffer stock from a levy on producers to absorb production in excess of quota. The agreement was renewed in 1975 with a higher price range. In 1980 the agreement was renewed with the quota system deleted.

The 1986 renewal added a mechanism for withholding sales as well as maintaining the buffer stock.

While in its early years the ICCA did not function because market forces kept the price high, since 1987 the price has fallen steadily to about one-half of the minimum intervention price. The buffer stock, now close to maximum level, could not be sold, except to dispose of beans which were deteriorating after lengthy storage. Failing to achieve the objective of stabilizing price, the buffer stock is being sold (Motluk, 1995).

A new agreement, negotiated in 1993, sets up a production management plan designed to improve the returns to growers.

Tea

Auctions

Up to 1939 most of the world's tea was sold by auction in London, while small quantities of tea from the then Dutch East Indies were auctioned in Amsterdam. The London auctions restarted in 1951. As tea production increased, but consumption by Western countries remained at around the same level, producing countries started local auctions. There are now tea auctions at regular intervals at Calcutta, Guwahati, Siliguri and Cochin in India, Colombo in Sri Lanka, Mombasa in Kenya, Limbe in Malawi and Jakarta in Indonesia. The quantities of tea sold at these local auctions has increased dramatically, while the quantity sold in London has decreased. Nevertheless, the London auction prices still set the price level for the world.

Tea produced in China is not sold by auction. This is sold by contract between the purchaser and the China national tea-selling organization.

International Tea Committee

The International Tea Committee (ITC) was set up in 1933 to monitor the first International Tea Agreement. Central to this task was the collection, tabulating and publishing of detailed statistics of all aspects of the worldwide tea trade. It has always had an office in London. The agreement terminated on the outbreak of the Second World War but the collection and publication of statistical data have continued to the present day. This is the sole purpose of the ITC; functions such as promotion, which the international organizations for the other crops (ICO and ICCO) carry out, are handled by separate bodies.

The Tea Council

This organization has an office in London. It provides information on various aspects of tea and organizes promotional activities. There is a medical consultant, who can give advice on medical matters. It has a small library.

RESEARCH

A substantial body of information on all three crops has been collected, and mostly published, in many countries. Individual organizations are too numerous to list, but they include universities, government research stations and units owned and operated by individual companies or associations of the particular industry in one country or group of countries. In all but some smaller countries, a body of local advice is available to growers. In many cases, this advice has been published in book form, and is often updated by regular publication of a journal.

4

BOTANY AND PLANT IMPROVEMENT

The coffee of world trade comes from two species, which developed in different regions of Africa. *Coffea arabica*, arabica coffee, forms the major proportion of world trade. *Coffea canephora*, robusta coffee, forms the remainder. Two other species, *Coffea liberica* and *Coffea excelsa*, produce inferior coffee and are only grown locally in some places in Africa, primarily for local use. There is no significant trade in these coffees. There are many other similar species within the genus *Coffea*. Many grow in very limited areas and there may be a few as yet undiscovered in the African rain forest. None of these has any commercial importance; some may have hybridized to some extent with *C. canephora*, of which there are several forms. Taxonomy within the genus is rather confused. Rakotomalala *et al.* (1993) investigated species differentiation, using the proportions of biochemical compounds in the species studied. Factors affecting the quality of the product have been discussed by Barel and Jacquet (1994). Clifford (1985) reviewed both chemical and physical factors affecting the quality of the final coffee product.

BOTANY

Arabica Coffee

C. arabica is the only tetraploid in the genus, with $2n = 44$. A few deviants have been found but are of little value. Arabica therefore does not hybridize with other species without special measures being taken. Two botanical varieties are known: var. *arabica* (syn. var. *typica*) and var. *bourbon*.

C. arabica is a small tree, which will grow unchecked to about 15 m in height; commercially the height is restricted by pruning (Fig. 4.1). Leaves are evergreen, glabrous and glossy with well-marked veins; at each node two leaves are formed, on opposite sides of the stems. The main stem is orthotropic; only one stem is formed unless it is cut to induce branching. Branching is

Fig. 4.1. Arabica coffee trees under light shade. These have been pruned to force the growth of several stems and to reduce the height. Papua New Guinea. (Photograph by K. Willson.)

dimorphic as there are two buds in each leaf axil. Normally the upper of these two buds grows from each axil to form a plagiotropic primary branch. The branches arise on the opposite side of each node, at 90° in var. *arabica* and 55° in var. *bourbon*. These plagiotropic stems bear the flowers and fruit, which are rare on orthotropic stems. The second bud will not replace a primary that falls off. If the main stem is cut, removing the influence of the apical bud, one or more new orthotropic stems arise from the second buds in leaf axils; repeated pruning can produce a multiple-stem tree. If the main stem is bent down, or a tree is planted at an angle to the vertical, new orthotropic stems will grow from buds along the upper surface of the main stem (Fig. 4.2).

There are six buds, one above the other, in each leaf axil on each plagiotropic stem. If conditions are right for flowering, it is usual for the first four buds to produce inflorescences. Cutting back a primary will induce secondary plagiotropic stems from one or more nodes immediately below the cut; this also occurs occasionally when not cut back. The lower nodes, below the secondaries, will produce inflorescences. The secondaries flower the year after their formation. Mature leaves are opposite and are dark green. Young leaves are lighter; in var. *arabica* there is a distinct bronzing, particularly at the tip. The leaves are elliptical, with an acuminate tip. Figure 4.3 shows a primary carrying flowers and fruit.

A seedling forms a short tap root, rarely longer than 45 cm, and lateral

Fig. 4.2. Robusta coffee seedling which was planted at an angle of about 45° with the ground. Note the strong growth of orthotrophic stems. (Photograph by K. Willson.)

Fig. 4.3. Primary plagiotropic branch of coffee carrying flowers and fruit, with one secondary branch.

roots branch off horizontally from the tap root. Four to eight of these turn to grow downwards for up to 3 m. The remainder form a mat of feeding roots. Roots will not go below the water-table – hence the importance of drainage.

Small, fragrant white flowers form in clusters in leaf axils. There can be up to 20 flowers in each axil (Fig. 4.4). Buds are initiated, grow to 4–5 mm in length and stop, suffering from water stress. Cannell (1985a) concluded that day length has little or no effect on flower initiation; in earlier years the effect of day length was believed to be significant. This stress is relieved following rain, and flowers open together 8–12 days after wetting, the number of days depending on the temperature. After opening early in the morning, the floral parts start to wither after 2 days and fall a few days later, leaving the ovaries to develop into fruit. Abnormal flowers – 'star flowers' – may form under adverse conditions, particularly high temperature. Petals remain small and stiff and often have a green tint. They do not set fruit. The fruit 'cherry' is a drupe (Fig. 4.5). In a normal fruit there are two seeds 8–13 mm long, with grooved flat surfaces against each other; the other surface is convex. Some fruit contain only one seed, which is known as a 'peaberry'. The fruit usually has a red skin, although yellow variants are known.

The cotyledons, which have a greenish tinge and are known as 'beans' or 'green beans' in the industry, are the parts required by the consumer. They are surrounded by a silvery testa, known as the 'silverskin', and enclosed in a parchment-like endocarp, known as the 'hull'. Beans from which the

Fig. 4.4. Coffee flowers (photograph from International Coffee Organization, London).

Fig. 4.5. Ripe coffee fruit (cherries) (photograph from International Coffee Organization, London).

endocarp and testa have not been removed are known as 'parchment coffee'. The mesocarp is a sweet pulp and is surrounded by the coloured exocarp (skin) (Fig. 4.6). Braham and Bressani (1979) reviewed information on the composition of pulp and parchment.

The cycles of both vegetative development and floral initiation are linked to the pattern of rainfall distribution. During dry and/or cool weather, floral initiation proceeds quickly, while vegetative growth is at a minimum. Moisture stress prevents the floral buds from developing into flowers. When the tree is rehydrated at the start of the wet season, the floral development restarts and vegetative buds produce new shoots. Fruits develop in the wet season and ripen in the following dry season. Drennan and Menzel (1994) reported that a period of water stress, caused by lack of rain, after earlier rain had rehydrated the tree, starting anthesis and vegetative growth, synchronized anthesis and accelerated vegetative growth.

C. arabica is self-fertile and seeds set even when the flower is bagged. The coffee of commerce consists of the dried seeds with the silverskin removed. There are many mutants and cultivars, of varying commercial importance.

The genetics of coffee have been reviewed by Carvalho *et al.* (1991), and Cannell (1985b) reviewed in detail the physiology of the production of crop by the coffee tree.

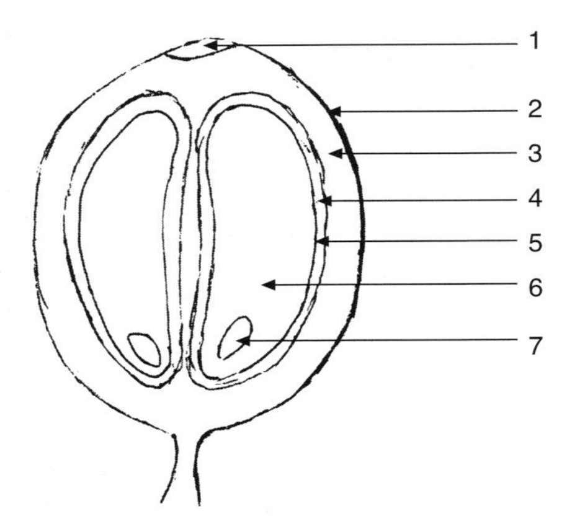

Fig. 4.6. Cross-section of coffee cherry. 1, Disc (navel); 2, exocarp (skin); 3, mesocarp (pulp); 4, endocarp (hull, parchment); 5, testa (silverskin); 6, cotyledon (bean); 7, embryo.

Robusta Coffee

C. canephora is similar in many ways to *C. arabica*, although it is diploid ($2n = 22$). It is a larger tree, with larger leaves, and the laminae between veins are more convex than in arabica (Figs 4.7 and 4.8). The taxonomy is confused and hybridization has probably occurred with other species of *Coffea* in the very large range of its natural habitat. It is self-sterile and cross-pollinates easily, so that characteristics quickly become disseminated in a large population. There are two types: Robusta is an upright tree whereas Nganda is a spreading form. Robusta is believed to be the species type and is therefore *C. canephora* var. *canephora*. Nganda has been named *C. canephora* var. *nganda* (Haarer, 1962a). Robusta seeds have a higher caffeine content (2–3%) than arabica seeds (1–1.5%) and are usually smaller; they have a lower quality than beans of arabica, lacking flavour and acidity.

Disease resistance

One important property of *C. canephora* is that most cultivars are resistant to coffee rust caused by *Hemileia vastatrix*. This resistance has been passed on to hybrids and some resistance to coffee berry disease (CBD) (*Colletotrichum kahawae*) has been found.

Fig. 4.7. Robusta coffee under coconut shade, Papua New Guinea (photograph by K. Willson).

Fig. 4.8. Primary branch of robusta coffee, from which a pair of secondaries is growing (photograph by K. Willson).

Biennial Bearing

A property which is of great practical importance is the strong tendency of coffee to bear biennially in plantation conditions. *C. arabica* suffers most in this respect. Biennial bearing leads to overbearing, which can put the tree under great stress so that it dies back, a result which, if severe, can kill the tree. Coffee has not developed the mechanism, common in many fruit trees, of shedding fruit that are beyond its capability to grow to maturity. Plantation conditions force flowering at a higher intensity than would occur in the area of origin, which leads to an insupportable crop of fruit. The burden of ripening a large crop restricts foliage production, so a year of low crop follows, which allows an excessive amount of foliage to form, and this, in turn, permits an excessively high number of flowers and hence crop. The alternation of high and low crops exhausts the trees, so pruning must aim to stabilize the amount of bearing wood each year. Gopal *et al.* (1993) discussed the reasons for the fluctuations in coffee yield.

PLANT IMPROVEMENT

Coffea arabica

The major proportion of research into the improvement of coffee plants has been carried out on *C. arabica*, whose product is of higher quality than that of *C. canephora* and therefore commands a higher price.

The earliest improvement work was directed towards improving yield and the stability of yield from one year to the next. A minor objective was to improve the adaptability to climatic variations, which could widen the range of environments where coffee could be grown commercially. Disease resistance was not important; leaf rust, *H. vastatrix*, had not spread to the major growing regions in South America. All the early plantings used trees which came from a narrow genetic base. Nevertheless, selection within the existing populations identified cultivars with superior characteristics of yield and quality. Several expeditions have collected coffee trees from the districts of origin of *C. arabica* in Ethiopia in order to widen the available genetic base. The improvement of *C. arabica* has been discussed in greater detail by Waiyaro (1983) and van der Vossen (1985).

The spread of leaf rust and the planting of arabica coffee in areas where leaf rust was active necessitated the extension of research into disease resistance and control. This started with the selection of resistant cultivars in South India. Concern about the potential of leaf rust to severely affect arabica coffee throughout the world led to the establishment of the Coffee Rust Research Centre (Centro di Investigacao des Ferrugens do Cafeeiro, CIFC) in Portugal, where there is no coffee industry to be affected by any pathogen which might escape. The importance of resistance to leaf rust increased when the disease arrived in South America, although this disease can be controlled by frequent application of copper fungicides.

The pathogen causing CBD, *C. kahawae*, spread from the indigenous species *Coffea eugenioides* in East Africa to all African coffee regions. Its further spread can be limited only by strict quarantine procedures. This disease can be controlled by frequent application of fungicides, which is expensive, or its severity reduced by manipulation of flowering to ensure that the climate is dry at the period of maximum susceptibility. Resistance to this disease is now an important objective of plant improvement.

C. canephora has genes that confer resistance to these diseases, so hybridization is a possible way to introduce resistance into *C. arabica*. This is exemplified by the natural hybrids Hibrido de Timor (*C. arabica* × *C. canephora*) and Congusta (*C. canephora* var. *ugandae* × *Coffea congensis*).

The early work on improvement of *C. arabica* by selection within existing populations identified a number of cultivars with superior yield characteristics, several of which have been planted on a large scale. The spread of CBD has greatly reduced the value of many of these.

The need to develop cultivars of *C. arabica* with resistance to the two major fungus diseases has led to the development of breeding programmes. These are mostly based upon crossing high-yielding cultivars in current use with various mutants of *C. arabica*, selections among the varieties brought back by collecting expeditions, and hybrids with *C. canephora* and other species of *Coffea*. The new cultivars produced by these programmes have to be assessed for a range of properties: resistance to leaf rust and CBD and possibly also to other pathogens of local importance, high yield, stability of yield from year to year and the quality of the coffee produced. A number of mutants grow to smaller trees than most common varieties. Dwarf trees are valuable as they can be planted more closely than trees of normal size. Dwarf or compact trees can also be allowed to grow without restriction for 2 years more than normal trees before they become so tall that they have to be pruned. Close planting increases the yield from a given area; closely planted trees of normal size can grow to impede or prevent access to individual trees. A selection of varieties used in breeding programmes is listed in Table 4.1. The list is not exhaustive; new discoveries and new hybrids are continually being incorporated into programmes, and others discarded.

Improvement of *C. arabica* can be divided into several stages, of which the first is selection and testing of possibly superior trees. Testing can identify the superior examples, which can be propagated by seed. Two cycles of selfing before testing will stabilize the characters of each line, then the superior varieties can be crossed one with another, in double or multiple crossings, and further improved trees can be selected within the progeny. Back-crossing may produce more improved varieties. Trees so produced will not reproduce exactly from their seed, so propagation must be vegetative, to produce both examples for testing and trees for field planting if the clone is released for large-scale planting.

Table 4.1. Cultivars used in breeding programmes (from Waiyaro, 1983; van der Vossen, 1985).

Name	Origin	Main characteristics
BA	India	Believed to be a natural hybrid of *C. arabica* and *C. liberica*. Resistant to most races of leaf rust
Blue Mountain	Jamaica, in Typica	Some resistance to CBD
Caturra	Brazil, in Bourbon	Dwarf
Cera	Brazil	Yellow endosperm
Congusta	Java, Indonesia	Natural hybrid of *C. canephora* var. *nganda* and *C. congensis*
Erecta	Puerto Rico, in Bourbon and Typica	Erect branching
Hibrido de Timor	Timor	Natural hybrid of *C. arabica* and *C. canephora*. Resistant to CBD and most leaf rust
K7	Kenya. Selected from Kents type (ex India)	Some resistance to leaf rust and CBD
Laurina	Brazil, in Bourbon	Narrow leaves and narrow, pointed seed. Low productivity and caffeine content
Maragogipe	Brazil, in Typica	Large leaves and seeds, long internodes, vigorous but low yielder
Mokka	Yemen	Compact conical tree with small leaves and round beans of high quality
Padang	Guatemala	High yield, partially resistant to CBD
Purpurescens	Several regions	Purple leaves
Pretoria	Guatemala	Large leaves on tall and vigorous treee. Very resistant to CBD
Rume Sudan	Sudan, adjacent to Ethiopia	Small, low-yielding, but a valuable source of resistance to CBD
Sao Bernardo	Brazil, in Typica	Compact dwarf
San Ramon	Brazil, in Typica	Dwarf, with very short internodes
SL28	Tanzania, in Bourbon	Commercial variety in Kenya. Normal size

Table 4.1. Continued

Name	Origin	Main characteristics
SL34	Kenya	Local selection; high yield, high quality. Planted commercially
S288 S333 S795	India	Believed to be natural hybrids of *C. arabica* and *C. liberica*. Resistant to most races of leaf rust

Two cycles of selfing of the clones will fix the major genetic characters. Crossing of pairs of these by hand pollination will produce F_1 hybrids, which can be propagated using the hybrid seed. Further cycles of selfing will produce varieties that are sufficiently stable to propagate from seed.

In vitro propagation by nodal cultures (Dublin, 1980) speeds up the cycles, which need plants that have been propagated vegetatively. The plants produced in this way have a long tap root, which is better than the several shorter roots produced by a cutting.

Coffea canephora

This species is self-incompatible. Each seedling is therefore a hybrid and it is not possible to find uniformity among plants from the same source. The wide geographical range of natural populations of *C. canephora* gives scope for widening the genetic base by crossing between separated populations. This could lead to greater variability between progeny and a wider scope for selection. The variability of open-pollinated seedlings means that superior individuals can be selected, but only vegetative propagation will produce uniform plants. When planting clonal orchards, it is important to ensure that the clones are cross-compatible so that pollination is adequate. Crossing of selected clones can produce some seedlings with improved characteristics. Van der Vossen (1985) includes a further brief discussion of the improvement of *C. canephora*.

Hybridization

Hybridization of *C. canephora* with *C. congensis* has produced the Congusta or Conuga hybrids in Indonesia. These are better adapted to higher and wetter regions than robusta. Direct hybrids between *C. arabica* and *C. canephora* are triploid and sterile, although Hibrido de Timor, discovered recently on the island of Timor, which is fertile and grown commercially, is believed to be a natural hybrid. In the hope of producing a hybrid that would grow in lowland Equatorial regions but produce coffee of quality approaching that of arabica, hybridization has been achieved after doubling the chromosome number of *C. canephora* by treatment of germinating seed or shoot tips with colchicine (Capot, 1972). The resultant hybrids are known as Arabusta. They have characters intermediate between the two parents, but rather closer to robusta. Superior individuals can be selected from these hybrids. They give better-quality coffee with a lower content of caffeine than *C. canephora* and can grow well in tropical lowlands. Back-crossing can produce some individuals with even better characteristics.

Doubling the chromosome number of triploid hybrids with colchicine produces hexaploids, which are fertile and similar in many ways to *C. arabica*. Little work has been done with these, as there are simpler ways to achieve the desired properties. Crossing tetraploid *C. liberica* with *C. arabica* produces hybrids known as Kawisaris. These also show sterility, to a degree which is very variable from plant to plant and year to year (le Pierres and Yapo, 1993).

Omondi and Owuor (1992) reported the performance of interspecific crosses and back-crosses involving arabica and tetraploid robusta species.

Karanja (1993) reported that the hybrid variety Ruiru 11, introduced by the Kenya Coffee Research Station has, since 1986, been planted in 36% of 140 estates questioned. Yields were 70–100% higher than with the old varieties which it replaced.

Breeding for resistance to coffee rust has been discussed by Carvalho *et al.* (1989). The dwarf variety Catimor is resistant to coffee rust. It has been developed from Hibrido de Timor and is now being planted in many regions. Van Boxtel (1994) studied the genetic transformation of coffee, using electroporation and biolistics to insert genes that impart resistance to diseases and pests.

5

Climatic Requirements, Soil Requirements and Management

CLIMATIC REQUIREMENTS

C. arabica (arabica coffee) developed in the highlands of Ethiopia. The climate is cool by tropical standards and rainfall fairly high, although the area is not as wet as in areas of dense tropical rain forest. The coffee became established as a midstorey tree in a semiopen forest, growing under at least partial shade. The soil had been leached by the rainfall but not so highly leached as the soil under dense rain forest on which tea developed. The soil on which arabica coffee developed is therefore slightly acid. The time of flowering became established so that the fruit develop in wet conditions and ripen in the sun in a dry period. Flowering had therefore to take place at the end of the previous year's dry period; the arrival of the rains triggers flowering (see Chapter 4, p. 37).

C. canephora (robusta coffee) developed in the lowland forest of the Congo river basin which extended up to Lake Victoria in Uganda. The altitude of this vast region varies from sea level up to about 1200 m in Uganda. The whole region has a high rainfall with a short dry season and was covered by dense rain forest. Temperature and humidity are always high, with only a little moderation for the altitude in Uganda. This coffee developed as a midstorey tree in tall, dense rain forest. The tree is therefore larger than arabica and has larger leaves, as it probably grew under conditions of deeper shade than arabica. Like arabica, it needed no mechanism for coping with drought conditions. Flowering had also to be arranged so that fruit ripened during the period of maximum sunlight, although this would be modified by shade from the tall forest trees. There are variants in robusta coffee which have developed in different parts of the region where conditions are slightly different. The soil of this area would be leached but there would have been, when undisturbed, a deep mulch layer of decomposed leaves and branches on the forest floor. This would probably be close to neutral in reaction and well supplied with available nutrients.

Climate

The water requirement of a crop is determined mainly by the amount of evapotranspiration. The amount of water that evaporates from an open water surface (E_o) can be calculated from meteorological data; plants transpire a lower amount than this (E_t) (Penman, 1948). The water transpired by plants is related to E_o by the 'crop factor' (E_t/E_o), which is usually between 0.85 and 0.95 for crops with a full ground cover of foliage. Measurements on arabica coffee led to a crop factor that was 0.5 in a dry month and 0.8 in a wet month (Wallis, 1963). These measurements were made on coffee with a wide spacing, so there was a lot of space between the trees. In dry weather, the soil surface was dry, without vegetative cover. Water loss from the soil was very small when the soil was bare of vegetation and covered with mulch. Summation over the year gave a total requirement of 950 mm, with monthly totals between 60 and 115 mm. The distribution of rainfall will never exactly fit the needs of the crop. Allowing for this, on soil with a reasonable water capacity, a minimum annual rainfall of 1300 mm can be postulated. The length of the dry season can be critical; the maximum period tolerable without rain is probably 4 months with favourable soil-water capacity.

Yacob-Edjamo *et al.* (1995) found that leaf rolling was a consequence of water stress, used by coffee to minimize water loss. Leaf dry weight and leaf water content were also significantly correlated with moisture stress.

Similar considerations apply to robusta coffee. Higher temperatures at the lower altitudes in which this species originated increase the transpiration rate. A necessary minimum annual rainfall of 1250 mm has been quoted, but this would require a very favourable distribution. A more realistic figure is 1550 mm. Some typical monthly rainfall statistics are given in Table 5.1.

Attempts have been made to relate crop yield to rainfall. Results for arabica cover such a wide range of variation that there can be no universal relationship. The rainfall must replenish the water used from the soil during the dry season and be sufficient to ensure full development of the fruit. A positive relationship between rainfall while crop is developing and crop yield has been found in El Salvador (de Castro, 1960), whilst a negative relationship between crop yield and the previous year's rainfall was reported from a very wet region in Costa Rica (Sylvain, 1959). In Papua New Guinea a negative relationship was found where excessive rainfall caused waterlogging (K. Willson, unpublished). Montagnon and Leroy (1993) studied the resistance of clones of *C. canephora* to drought in West Africa.

Rainfall distribution is important for coffee in that change in internal water tension triggers flowering. The flowering follows the first rain at the end of a dry period. Where rainfall is unimodal, in regions furthest from the Equator, there is only one flowering; in Equatorial regions, where rainfall is bimodal, the coffee flowers twice each year. The climatic regime for coffee must include at least one season dry enough to induce water stress in the plants.

Table 5.1. Typical monthly rainfall on coffee.

Country	Altitude (m)	Species	Rainfall (mm)												
			Jan.	Feb.	Mar.	Apr.	May	June	July	Aug.	Sept.	Oct.	Nov.	Dec.	Total
[1]Papua New Guinea	1575	Arabica	247	247	258	168	123	85	92	114	119	156	177	232	2018
[2]Venezuela	880	Arabica	87	86	81	153	131	88	73	93	117	202	218	152	1481
	1200	Arabica	56	51	60	122	132	167	141	126	114	131	120	70	1290
[3]Tanzania	1333	Arabica	28	48	121	493	453	113	49	29	37	35	55	26	1487
	1455	Arabica	58	97	119	655	558	213	127	58	25	46	58	69	2083
[1]Côte d'Ivoire	150	Robusta	40	8	265	205	300	620	260	160	450	285	140	5	2738

Data from: 1, K. Willson, unpublished; 2, M.N. Clifford, 1995, personal communication; 3, Haarer, 1962b.

The optimum temperature range for arabica coffee is 15–24°C, which reflects its origin in a high-altitude Equatorial region with some shade. Over 25°C, the photosynthetic activity is reduced. Continuous periods over 30°C damage the leaves. Frost destroys leaves and fruit. Sharp daily changes between low and high temperatures by night and day damage the plants; this is known as 'hot-and-cold disease' (Chapter 8, p. 97). Leaves discolour to white or yellow, initially on the margins. Leaves distort, do not grow to full size and may eventually scorch and fall. In extreme cases, excessive formation of secondary and tertiary stems may occur and shoot tips blacken, distort and shrivel.

For robusta coffee, optimum temperatures are higher, 24–30°C. Foliage is destroyed at 5°C, but will withstand an occasional fall to 7°C. Long periods at 15°C are harmful.

Atmospheric humidity has a major effect on evapotranspiration. Moisture loss is reduced as humidity rises. Cloud cover and mist, to a greater extent, increase humidity. Therefore, in regions which have a lot of cloud during the dry season and regions where misty conditions last for many days, coffee can be grown satisfactorily with a lower rainfall. Mist often results in condensation on trees; this drops to the ground and effectively increases the rainfall.

Wind can have significant effects. In the first instance, wind increases evapotranspiration and therefore rainfall requirement and/or moisture stress in the trees. Cold wind can intensify the problem of hot-and-cold disease, while hot wind can scorch the more exposed foliage. In either case, crop yield will be reduced. Where strong winds are frequent, wind-breaks need to be provided.

SOIL REQUIREMENTS

An important requirement for coffee is good drainage to enable excess rainfall in the wet season to drain easily away from the roots; waterlogging destroys the roots and therefore hinders subsequent water uptake.

Soil should have an open texture to facilitate drainage and assist the roots to develop. Roots need to go deep enough to be able to extract water from a large volume of soil during the dry season. Roots were found to explore actively up to 3 m depth of soil in Kenya (Nutman, 1933). At the end of the dry season, the soil is often at its wilting point down to the 3 m level (Blore, 1966). There should be no physical obstructions, such as rocks, hardpans and heavy clay, to restrict root growth downwards, as the amount of water available to the plants would be reduced. Moisture stress may therefore reduce yield if rainfall is below average in the wet season. Above-average rainfall will have a similar effect, as excess water will only drain away slowly and the lower roots will become waterlogged.

Chemically the soil should preferably be slightly acid, with a pH between 5.2 and 6.3 for both arabica and robusta. In practice, coffee is grown on more acid soils (pH down to below 4.0) and on alkaline soils (pH up to 8.0). In such cases, mineral imbalances will occur and yield will be restricted. Soil management must aim to amend the soil to give a pH within the optimum range. Levels of pH and the contents of the major base nutrients of some soils on which coffee is grown are listed in Table 5.2. The wide range of pH, 4.3–8.0, will be noted, and also the effects of mulching and burning of vegetation after clearance, which raise the pH considerably.

The data in Table 5.2 show that levels of available phosphorus are low, except for topsoil in 30-year-old coffee where application of phosphate fertilizer over many years has built up a significant amount, despite a low pH value. The levels of potassium in the soils are the lowest of the three bases. Heavy cropping over a period of years will reduce the level of potassium, and the pH will fall. Application of potash fertilizer is vital to keep soil levels up and to prevent potassium deficiency from restricting crop. The optimum level of fertilizer application will increase as the pH and the amount of potassium available in the soil fall. The level of calcium in the soil is usually adequate for the coffee, except perhaps when the pH becomes very low. Magnesium can be in short supply more frequently than calcium and application of magnesium fertilizer becomes necessary. Very acid soils are often treated with lime; this will increase the calcium content. If dolomitic limestone is used, the magnesium content will also be increased.

Mulching is common practice in coffee. The nutrients in the mulch will become incorporated in the soil as the mulch decomposes. If the mulch has bases in proportions different from those of coffee or from those desirable in coffee soil, the levels in the soil will change. This can have adverse effects on coffee nutrition where mulch is applied regularly over a long period of time (Mehlich, 1965). For example, elephant or napier grass (*Pennisetum purpureum*) is often used. This has a high potassium content so that the level of potassium in the soil increases to the extent that magnesium deficiency is induced. Sisal (*Agave sisalana*) has a high calcium content. The waste from sisal processing is used as a mulch on coffee where it is available. Regular mulching increases the calcium in the soil to a level where it interferes with the uptake of potassium and magnesium.

SOIL MANAGEMENT

Land clearance and preparation must be carried out so that the soil is suitable for coffee and any loss of topsoil is kept to a minimum. The method used will depend on the vegetation on the site.

Where the site is forested, large trees should be killed slowly before clearance starts. Harmful root diseases, such as *Armillaria mellea*, which may be living on tree roots will die along with the tree if the process is slow. The risk of these diseases invading the new coffee is therefore minimized. Trees should be

Table 5.2. Chemical analyses of soils under coffee.

Country	Coffee		Depth of sample (cm)	pH	P ($\mu g\ ml^{-1}$)	K (meq. %)	Ca (meq. %)	Mg (meq. %)	CEC (meq. %)	
	Species	Age								
[1]Papua New Guinea	Arabica	30	0–8	4.3	69	0.2	0.5	0.4	20	
			8–24	4.4	12	0.3	0.3	0.2	15	
			24–60	4.9	2	0.1	2.3	1.4	10	
			60–120	5.4	1	0.4	5.2	3.3	12	
[2]Kenya	Arabica	5	0–8	5.4	15	1.5	7.0	3.8		Not mulched
			0–8	6.7	40	3.1	7.8	5.7		Mulched
[1]Papua New Guinea	Robusta	3	0–8	6.4	6	0.5	18.0	5.9	29	
[3]Central African Republic	Robusta	–	0–10	5.7		0.4	4.4	0.7		Ready for planting; no burning
			0–10	8.0		0.8	9.5	1.2		Ready for planting; vegetation burned

P, phosphorus; K, potassium; Ca, calcium; Mg, magnesium; meq., milliequivalent; CEC, cation exchange capacity.
Data from: 1, K. Willson, unpublished records; 2, Robinson and Hosegood, 1963; 3, Deuss, 1969.

killed by ring-barking or frilling; this treatment should be inspected regularly as bark often grows across the cut to revive the tree. Dead trees are felled and removed from the site. Smaller vegetation must be cut and can be stacked for burning. Ash from fires must be spread over the field to avoid high localized concentrations, which can produce undesirably high concentrations of bases. It may be convenient to pile the cut vegetation in lines between the intended lines of coffee. Coffee will then not be planted into ashes.

Where vegetation is much lighter, such as savannah or grassland, it will be sufficient to knock down trees and burn the whole area. After burning, the field should be ploughed as deeply as possible to incorporate the ash.

On sloping land in areas where heavy rainfall can cause significant soil erosion, measures must be taken to limit this. Gently sloping terraces leading to drains running downhill are best where rainfall is heavy. Drains on the contour without outlets to downhill drains are adequate where rainfall is lower. They have the advantage that the water will seep into the soil and be retained for the coffee.

On flat land where rainfall is heavy and in any situation where the water-table can rise above 3 m below the surface, deep drains should be made to take water away. This is particularly important where flooding may occur or on drained swampland. Where the soil does not drain freely, subsidiary shallower ditches should be dug between coffee lines to take water to the main drains. These may need to be as close as between alternate lines of coffee.

A cover crop should be sown as soon as possible after ploughing and draining. This may be a short-lived crop, such as oats (*Avena sativa*) or buckwheat (*Fagopyrum esculentum*), or a perennial, which may remain as a cover crop under the coffee. Spreading legumes are very suitable for this purpose. The cover crop will minimize soil movement and help to establish a surface mulch or an inter-row crop before the coffee is planted.

ENVIRONMENTAL IMPACT OF COFFEE GROWING

Interest in the environmental impact of coffee growing and processing is increasing. In the field, there is concern over deterioration of land planted with poorly managed coffee, and interest in 'organic' coffee is increasing. The environmental effects of effluents from coffee production are also giving concern. The International Coffee Organization hosted a 2-day seminar on these subjects (International Coffee Organization, 1996c).

6

FIELD MANAGEMENT

SITE SELECTION

Soil and climate must be considered, along with several other factors, when selecting a site for planting coffee. For a fuller discussion see Chapter 5.

The soil should be deep and free-draining. Shallow soils restrict the capability of the plants to withstand periods of dry weather. Such soils can only be planted where either the dry season is short or irrigation will be provided. Chemically the soil should be slightly acid. Very acid soil will not kill coffee, but fertilizer needs, particularly for potash, will then be greater than on less acid soil. Soil that is neutral or alkaline can be acidified by incorporation of sulphur (see Chapter 16, p. 197). Soil that is largely limestone should be avoided, as excessive quantities of calcium will hinder the uptake of magnesium and possibly potassium.

Climatic requirements have been discussed in Chapter 5. It is important to know whether the rainfall is unimodal or bimodal. A bimodal rainfall will initiate two flowerings and hence two peak periods of harvest. The total period of harvesting during the year is likely to be longer, so that the daily amount which has to be harvested and processed will have a lower maximum. This has implications for harvesting labour requirements and maximum factory capacity for wet-process coffee.

The aspect of the site may be important. South-facing slopes in the northern hemisphere and the reverse further south will receive more sun and therefore be warmer than those facing in the opposite direction. This factor can be important in marginal areas. Valley bottoms may be prone to frost in marginal areas, but slopes will permit cold air to flow away from coffee. Cool night conditions alternating with high day temperatures can cause hot-and-cold disease.

Wind can be troublesome. Shade trees may reduce the impact of wind on coffee but wind can funnel under shade trees and have the reverse effect. Lines of close-planted trees can form wind-breaks but are best planted some distance from the coffee to minimize competition.

A permanent, reliable source of clean water is vital. Apart from personal needs, processing by the wet method needs a continuous water flow. Water will be required for a nursery and for irrigation if this is to be installed. Processing is often at its peak in the dry season, at which time irrigation water, even if only for a nursery, will also be required.

Access to and within the site is important. Personnel, supplies and products must be moved easily in and out. Morris (1997) discussed design, construction and maintenance of estate roads.

Coffee has to be processed on or close to the plantation, so a suitable piece of land must be available. It does not necessarily need to be flat; wet processing uses a downward flow of water, so a slope rising from a flat area is very convenient. The flat area must be large enough for fermenting tanks and sun-drying equipment.

Although many coffee factories include hot-air drying equipment, the need for fuel is not very large. The need for hot-air drying is intermittent; fuel oil is the most convenient source of heat. There is no large requirement for firewood.

CHOICE OF CULTIVAR

The choice of species, arabica or robusta, will be dictated by climatic conditions. It is quite probable that the species to be grown will be chosen first and then a suitable site. In large areas, even whole countries, there is no choice, as suitable conditions for only one or the other exist. In some areas, the climate is such that both species can be grown. Ramachandran *et al.* (1993a) reported that *C. arabica* cv. Cauvery could be grafted on to *C. canephora* cv. 5274. The two cultivars cropped at different times, permitting each to be harvested separately.

The heterogeneous nature of robusta arising from self-sterility means that choice is limited to seed from one or more existing plantations or seed orchards. A choice can be made between the upright (var. *robusta*) and spreading (var. *nganda*) forms. The growth habit will control the choice of field spacing and pruning systems. The decision may follow local custom, particularly where the growers are smallholders. Planting of tested varieties propagated vegetatively is becoming more popular; it is standard practice on estates in Indonesia and increasingly followed by smallholders in Cameroon, Côte d'Ivoire and Uganda. The planting of hybrids, such as Arabusta, which must be propagated vegetatively, is an alternative in the few countries where such material is available.

Many cultivars of arabica are available. Some are universal, while others have been developed locally, specifically for local conditions. All cultivars available are believed to have come from one of the two main varieties, var. *arabica* and var. *bourbon*. Var. *arabica* and its derivatives are the most widely planted at present. These are generally vigorous but can be outyielded by var. *bourbon* under favourable conditions. Var. *bourbon* has a greater resistance to coffee berry disease, so cultivars derived from it are being more widely planted in areas where this disease is prevalent. Among the known cultivars are at least one that is typically larger than trees of the original variety and several dwarf varieties. With spacing and pruning techniques adjusted for the size of the trees, dwarf varieties are being planted in an increasing proportion of new plantings.

Some of the more widely known cultivars are listed in Table 4.1 (p. 42). Expeditions to the regions where arabica coffee originated and breeding programmes in many research centres are producing new varieties, most of which are hybrids. In any growing area, there will be advice on suitability and availability of the most appropriate cultivars.

PROPAGATION

Seed

This traditional method is the normal for arabica as its self-fertility ensures that the progeny are true to the cultivar. It is not suitable for hybrids or for the selected robusta material, which must be propagated vegetatively. Seed should be collected from selected trees and then pulped and washed but not dried, as wet seeds retain greater germinative power. Van der Vossen (1979) reported that the germinative power of seeds was maintained most effectively (up to 30 months) in seeds containing 41% water, stored at 15°C.

Seed can be pregerminated, but this stage is not essential. For germination, seeds should be planted into a bed of soil of fine tilth, 1–1.5 cm deep, at a spacing of 2 cm × 2 cm. The bed should be mulched and shaded from strong sunlight. Andrade (1988a) recommends soil in a bed made from concrete blocks 40 cm × 20 cm × 20 cm. Using 1 kg of seeds (2500–3000 seeds) m^{-2} gives a spacing approximately 2 cm × 2 cm. After placing the seeds on the soil surface, they should be pressed into the soil by pressure from a flat hand. Andrade (1988a) recommends covering the seeds with a 1 cm depth of sand, which should be removed after 5 days. Alternatively, seeds can be germinated spread on sacking or black plastic covered with damp sacking. Another method is to mix seed with damp vermiculite (three parts seed to one of vermiculite) in a plastic bag. Seeds germinated in soil are transplanted when the two cotyledon leaves are fully developed – after about 10 days in a warm district or up to 15 days where it is cooler (Andrade, 1988a). Germinated in other ways, they are transplanted when the hypocotyl has emerged. Seeds are commonly sown thickly in soil germination beds and transplanted at 'soldier' stage, i.e. about 2 cm tall with the cotyledons still enclosed in the 'parchment' skin. Pregermination enables poor seeds to be discarded.

Seedlings can be raised in beds, polybags or plastic sleeves. Beds should be raised above ground level to prevent waterlogging and the soil should have been dug to a depth of 1 m. It helps to incorporate coffee pulp or well-rotted manure or compost in the soil, or other fertilizers as may be found necessary. Andrade (1988a) recommends 2.5 kg of diammonium phosphate and 2.5 kg of magnesium sulphate 1000 kg^{-1} of soil. Plastic sleeves should be 9 cm diameter × 35 cm deep for robusta (Forestier, 1969a) and 15 cm diameter × 20 cm deep for arabica (Andrade, 1988a). Spacings in beds for arabica should be 12–20 cm × 25 cm. Paths between beds give access and shade must be provided, either individual shade up to 1 m above each level or high shade at a height of about 2 m over the nursery (Robinson, 1964) (Fig. 6.1).

The beds must be kept moist; a grass mulch is helpful in reducing moisture loss from the soil. Shade should be reduced slowly as the plants grow; plants should have been in full light for several weeks before they are taken to the field.

Andrade (1988a) recommended that, in order to reduce osmotic pressure and transpiration loss, before moving plants from the nursery to the field for planting they should be sprayed with a sugar solution (10 kg 100 l^{-1} water plus 25 ml of a 'sticker') on each of the 4 days prior to lifting from the nursery. As the plants are lifted, the roots should be trimmed to two-thirds of their length. This stimulates formation of a dense root system.

Fig. 6.1. Nursery of sleeved robusta coffee plants, ready for planting out, Papua New Guinea (photograph from W.R. Carpenter & Co. Ltd, Papua New Guinea).

Nursery beds should not be used for more than two consecutive lots of seedlings. Soil fertility is reduced, while pests and diseases will multiply.

The use of synthetic seeds has been suggested by Toruan-Mathis and Sumaryono (1994).

Vegetative Propagation

Cuttings

Multinode and single-node cuttings have rooted successfully. The fourth to sixth leaf pairs from active orthotropic stems are said to give the highest success rate (Forestier, 1969b). In suitable conditions, either one leaf of a pair on single-node cuttings can be removed or the stem can be cut down the centre to give two single-leaf cuttings. The leaves are often cut to one half of their length. Cuttings are placed in small individual sleeves containing an infertile rooting medium, such as subsoil or a sand–peat mixture. Forestier (1969b) recommends well-rotted sawdust. A rooting hormone is not always recommended. Andrade (1988a) recommended a mixture of three parts of crushed cocoa husk or well-washed bark and one part of sand. Cuttings should consist of a single node cut from a stem of diameter 5–8 mm. One leaf of each pair is removed, and the other is cut to one-half of its length; the stem on each cutting should be 4–6 cm in length. The lowest 2 cm of each cutting should be immersed in a solution of a rooting hormone for 15 s and put immediately into the rooting medium. The beds of cuttings are sealed under a polythene tent after watering unless they can be misted; either method must be under a high shade. As the cuttings grow, the tent is opened in stages. Rooting will take up to 3 months (Fig. 6.2), after which the plantlets are transferred to larger sleeves (Andrade (1988a) recommended 15 cm diameter × 20 cm height) containing a good forest soil or a mixture of sand and rich compost. The shade is then thinned slowly. Plants should be in full sunlight for several weeks before planting out. Wamatu (1990) discussed the vegetative propagation of the disease-resistant cultivar Ruiru 11.

Budding or grafting

This method was used in Java before 1900. Success rates reported in more recent trials vary. A range of standard methods have been tried. Before any large-scale use of these methods could be undertaken, the compatibility of rootstock and scion would have to be checked and the most successful technique determined.

Le Pierres (1987) pointed out that compatibility between species varies from one continent to another. He suggests using an interstock, such as *Coffea congensis* or *Coffea liberica*, between *Coffea canephora* and *Coffea arabica*. Kumar and Samuel (1990) suggested grafting on to nematode-resistant stocks as the best way to defeat nematode problems.

Fig. 6.2. Rooted robusta coffee cuttings, Côte d'Ivoire (photograph by K. Willson).

Andrade (1988a) described grafting in the nursery. Wedge-cleft grafting is recommended by Ramachandran *et al.* (1993b) for grafting an arabica cultivar on to a robusta tree.

Propagation *in vitro*

Propagation of some genotypes of coffee by tissue culture has been achieved; the methods and conditions must be carefully evaluated and controlled. Bieysse *et al.* (1993) reported the conditions under which four out of seven genotypes produced somatic embryoids that could develop into plantlets. Marques (1993) reported somatic embryogenesis of *Coffea eugenioides*, a species with potential for introducing genetic variability. Kahia (1993) reported the production of plantlets from *C. arabica* shoot tips.

Microcuttings can be produced by treating axillary buds with cytokinins. Orthotropic shoots are produced; these are subdivided to give a succession of cuttings (Berthouly *et al.*, 1987). Multiplication on a semisolid medium is slow; a much higher rate is obtained by temporary immersion, using the apparatus devised by Teisson (1994).

Somatic embryogenesis of coffee has been achieved on a semisolid medium (Starisky, 1970) and a liquid medium (Berthouly, 1991). Embryos present in the callus formed will develop when treated in the temporary-immersion apparatus.

FIELD PLANTING

Spacing

For maximum use of sunlight, a complete cover of vegetation is desirable, particularly in perennial crops. However, in coffee, other factors are important, so that a complete cover by close-planted trees is unusual. Sunlight on the sides of trees stimulates ripening of the cherries. Harvesting needs space to move around and between trees, while mechanized soil-maintenance operations need space between trees. Soil exposed to sunlight will be hotter than under shade so can lose more moisture, but mulching will reduce the loss.

Close spacings give higher yields when trees are small but the advantage reduces as trees mature, and the yield from close spacings will fall if the trees are too close and interlock or shade each other. The yield from closer spacings falls when the root systems of adjacent trees overlap. Browning and Fisher (1976) showed that yields are highest from an optimum density and fall at higher densities (Fig. 6.3). Pruning techniques determine the size and shape of trees. This interacts with spacing. Edjamo *et al.* (1993) investigated the relationship between canopy shape, number of bearing heads (stems carrying crop in multiple-stem coffee), density of planting and yield. They found that, as the number of bearing heads increased, the yield per head fell but total yield increased. Castillo *et al.* (1995) showed that leaf number and leaf area decreased but leaf-area index increased as a quadratic function of planting density. Light interception was not changed by the spatial plant arrangement. The upper part of the trees intercepted 73–78% of the incident light.

Traditionally, arabica coffee is planted at a density around 1350 trees ha^{-1}, typically at 2.7 m apart in square or triangular arrangement or in lines 3.0 m apart, with 1.5 m, 2.0 m or 2.4 m between trees in the lines. Dwarf cultivars would be planted more closely, around 2.0 m × 2.0 m or 1.5 m.

Robusta coffee would be planted at a lower density, little more than 1000 trees ha^{-1}. In regions where growth is vigorous, 900 trees ha^{-1} is adequate. A typical arrangement would be 2.25 m between trees in lines 4.0 m apart; this gives 1111 trees ha^{-1}. Modern practice is to increase the planting density to around 2000 trees ha^{-1}, with typical arrangements 3.0 m × 1.7 m and 2.8 m × 1.8 m.

Planting in the Field

If a cover crop has been planted at the end of the land preparation, this must be removed from the soil around the planting holes. It can be dug out around each tree site or ploughed in along each line of coffee, leaving a wide band midway between each line. Tree sites should be marked by stakes.

Planting early in the wet season gives the plants a long period of adequate

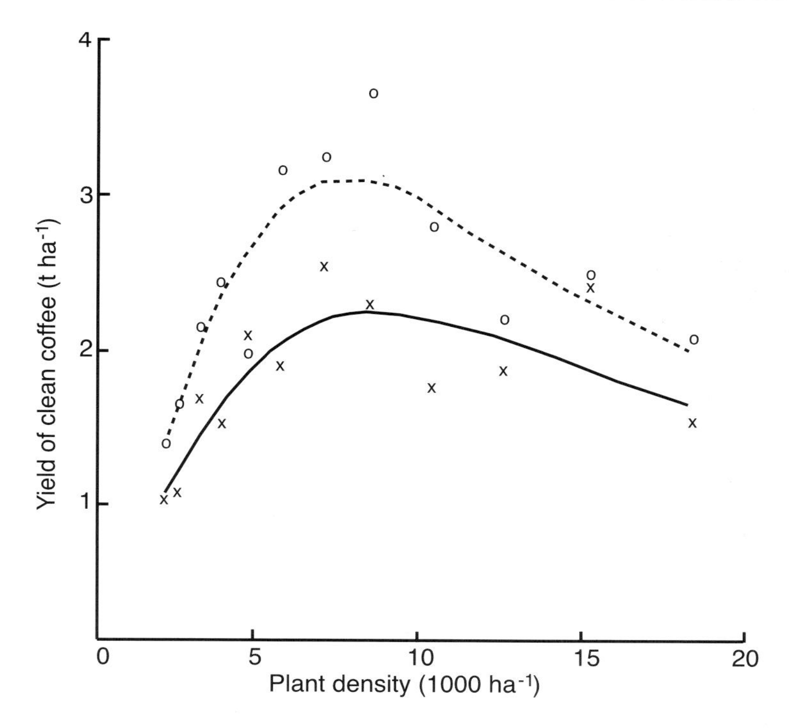

Fig. 6.3. Relationship between density of planting of coffee plants and annual yield of clean coffee. The lines represent curves fitted to the data marked by circles and crosses, respectively. Data for arabica coffee, cultivar SL28, at the Coffee Research Station, Ruiru, Kenya, for 1973 (circles and broken line) and 1974 (crosses and solid line). (From Browning and Fisher, 1976.)

moisture and low evapotranspiration loss to become fully established. The later the planting, the greater the risk of loss in the dry season. The availability of irrigation makes the choice of planting time much more flexible. Planting can then be done in the dry season, but plants must be shaded against hot, sunny conditions. In some areas, it is desirable to dig planting holes several months before planting to allow the soil to weather. They will be reopened shortly before planting. Fertilizer or organic manure should be mixed with the soil when the holes are dug or reopened.

Plants from a nursery bed can be taken to the field with a ball of soil attached if the soil will remain in place. Otherwise, they will arrive with bare roots. Plants in transit must be kept shaded and moist, preferably under wet sacking. Plants in polybags will retain their soil; the bag should be cut off when the plant is in position over the hole. Care must be taken that every bag is removed. Bags left in position greatly restrict root development and plants may die in dry weather. Each plant should be placed in its hole and the soil firmed around so that the soil level is at the same place on the stem as it was in the nursery. The soil around the stem should be higher than the surrounding field, so that after the inevitable settlement it is not possible for a pool of water to collect around the stem. The stake used to mark the plant position should be placed at an angle over the new plant as a marker and temporary shade put in position. If mulching is standard practice, this can be laid close to the young plants. A cover crop will regrow around the new plants but must be prevented from coming too close; leave a circle of 10 cm radius clear round each plant. If cutworm or any other pest is known to attack young plants, an appropriate insecticide can be applied around each plant.

Plants are normally set vertically, but planting at an angle of about 30° to the ground has been practised. This saves either cutting or bending the stem to initiate the growth of several stems to form a multiple-stem tree (see Fig. 4.2, p. 35). The root system is inevitably not as deep as when planted vertically and is to one side of the tree, not directly under and around the stem. Trees are much more likely to fall over, often by rotation around the axis of the stem. Snoeck (1963) showed that trees planted at an angle produced fewer stems than trees planted vertically followed by bending the stem over when it is about 1 m long.

At times, coffee seeds are planted directly into the field: ‘seed-at-stake’. While this eliminates the need for a nursery, maintenance of the young trees is difficult. They can easily be smothered by weeds or damaged during weeding. It is not possible to select good plants or discard poor ones without leaving a vacancy. A modified system is used in Brazil; five or six seeds are planted at each site. Losing one seed does not create a vacancy and multiple-stem plants are formed immediately. Each multiple plant is known as a ‘Cova’.

Shade

The natural habitat of most coffee species is the understorey of a forest. Therefore early coffee plantings were under shade. The early practice of planting under forest trees remaining after clearance of the lower vegetation was replaced by planting under specially planted trees. It was found later that, in many situations, coffee would grow well without shade and outyield shaded coffee (Triana, 1957) (Fig. 6.4). The higher yields needed more fertilizer. Shade stimulates early flowering (Perkins, 1947; quoted by Kimemia and Njoroge, 1988) (Fig. 6.5). In general unfertilized shaded coffee yields a little less than unfertilized unshaded coffee, but fertilized unshaded coffee yields much more than fertilized shaded coffee. Fournier (1988) reviewed coffee planting under shade and full sunlight. While unshaded coffee gave yields 10–20% higher than shaded coffee, he suggested that a final decision be deferred until more data are available. The long-term effect of higher yields from unshaded coffee

Fig. 6.4. Yield of coffee from the two most common varieties, shaded and unshaded, with and without fertilizer (from Triana, 1957).

should be evaluated before a final recommendation is made. Figure 6.6 shows young coffee plants under temporary shade.

Andrade (1988b) discussed the value of shade and the distribution of shade trees in the plantation. Plantations requiring high yields from a system with high inputs do not usually shade the coffee. Smallholders, however, who are restricted to limited inputs, retain shade, often as part of a mixed cropping system.

Shade has a number of effects on the plantation system in addition to reducing the amount of light reaching the coffee trees. Shade-tree roots often explore a different level of soil from coffee roots and absorb nutrients that ultimately reach the coffee as leaf fall. On the other hand, they compete for moisture, although the coffee may use slightly less because the shade reduces the temperature, wind speed and insolation around the coffee level. Other potential effects of shade include: incidence of pests and diseases, expenditure on shade-tree maintenance, products (timber or fruit) from trees, damage to

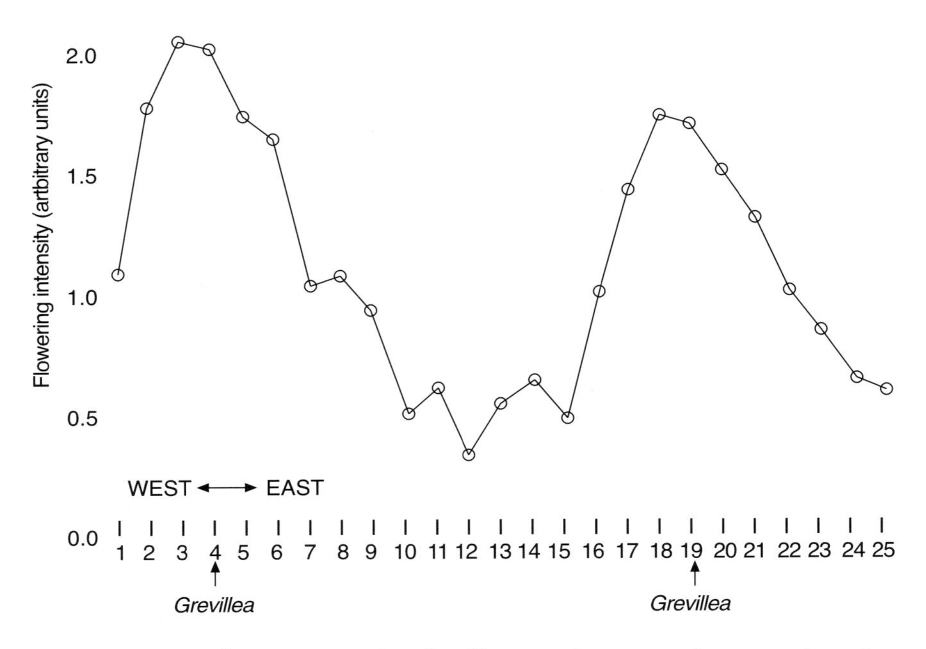

Fig. 6.5. Average flowering intensity of coffee trees in consecutive rows, planted with *Grevillea robusta* shade trees every 1.5 coffee rows (from Kimemia and Njoroge, 1988).

Fig. 6.6. Young arabica coffee under temporary leguminous shade, Papua New Guinea (photograph by K. Willson).

coffee by work on shade trees and obstruction of machinery. The level of such effects is variable. Local conditions must be considered when deciding on the need for shade.

Many species of tree are, or have been, used as shade. Many, used mostly with arabica coffee, are leguminous: *Acacia* spp., *Albizzia* spp., *Cassia* spp., *Erythrina* spp., *Gliricidia* spp., *Inga* spp. and *Leucaena* spp. (see Fig. 4.1, p. 34) are commonly used. *Grevillea robusta* is a good shade tree, although it is not a legume. *Casuarina* spp., a nitrogen-fixer but not a legume, is used but is not ideal. Kumar *et al.* (1992) discussed the use of *Parkia roxburghii*. Somarriba (1992) reported that shade trees which can be felled for timber (*Cordia alliodora*) increased the total return from the plantation. Robusta coffee has been tried under several tropical lowland tree crops. Oil-palm and *Hevea* rubber create too much shade, while bananas compete excessively for water and nutrients. Robusta coffee under coconuts is a successful and widely used system. Robusta can also be successful under *Cola* spp. Shade trees should be established before coffee is planted. Kimemia and Njoroge (1988) reviewed the effect of shade on coffee.

Many smallholders plant a multicrop system, which often includes coffee as a midstorey tree. Figure 6.7 shows arabica planted in traditional smallholder fashion, planted irregularly and mixed with bananas and other trees. Kurikanthimath *et al.* (1994) described a multistorey cropping system using coffee, clove (*Syzygium aromaticum*) and pepper (*Piper*

Fig. 6.7. Robusta coffee planted as a midstorey tree, typical of smallholder plantings, Cameroon (photograph by K. Willson).

nigrum). Herzog (1994) reported that smallholders in Côte d'Ivoire plant coffee under a mixed stand of shade trees, most of which are used traditionally for a variety of purposes.

Cover Crops and Intercropping

Wide spaces between coffee trees can suffer from erosion and also foster many weeds. A strong cover crop can minimize erosion and discourage weeds. If it is a legume, it will add nitrogen to the soil. Tall grasses and erect legumes can provide shelter for young coffee trees. A tall, short-lived legume, such as *Flemingia* spp., *Sesbania* spp. or *Tephrosia* spp., gives ground cover and shelter and lives long enough to shade young coffee. Kiara and Naged (1995b) found that there was no advantage from planting Catimor in temporary shade. Shade every two coffee rows adversely affected the growth of young coffee.

Cereals, such as buckwheat (*Fagopyrum esculentum*), can be used in arabica coffee. Guatemala grass (*Tripsacum laxum*) provides cover and shelter but must ultimately be removed entirely. The pasture legumes, such as *Desmodium* spp., *Calopogonium* spp., *Pueraria* spp. and *Stylosanthes* spp., are suitable in appropriate locations. They must be kept away from the coffee but can be retained along the centre of the interlines. Snoeck *et al.* (1994) reported that cut grass had been used traditionally in Burundi, where coffee is planted on steep slopes and there is a long dry period. Grass is becoming more difficult to obtain. Legumes planted as an alternative do not grow well without application of phosphate fertilizer and dolomite.

Short-lived crops can be grown along the interlines. Smallholders in suitable regions include coffee in multiple-cropping system; groundnuts (*Arachis hypogaea*), soybean (*Glycine max*), sweet potato (*Ipomoea batatas*) and vegetables are often planted. Njoroge *et al.* (1993) showed that it was profitable to grow potatoes, tomatoes or *Phaseolus vulgaris* within a coffee plantation.

Wind-breaks

In exposed situations and in areas subject to frequent strong winds, it is desirable to plant wind-breaks. Lines of trees at right angles to the wind direction are effective, but the trees should not be too dense. The air needs to filter through the canopy; if it is forced over the trees it creates turbulence beyond it, which can be very damaging.

Lines of trees within the field must be compatible with the coffee; species used as shade trees can be planted closely in a line. Outside the fields, taller trees, such as *Eucalyptus* spp. and *Pinus* spp., can be used.

Mulch

Mulch can play an important part in coffee husbandry, especially in dry areas. It helps to conserve moisture in the soil, minimize erosion and discourage weeds. As it breaks down, it releases nutrients into the soil and increases the content of organic matter. Many experiments have demonstrated that mulch increases the yield of coffee (e.g. Robinson and Hosegood, 1963). Any easily available foliage can be used as a mulch. Natural savannah grass, elephant or napier grass (*Pennisetum purpureum*), guatemala grass (*T. laxum*), maize stover (stalks and leaves), banana trash, sisal waste and a variety of legumes have all been used.

The disadvantage of mulch it that it is expensive; often it has to be cut, and always it must be transported and spread. Under some circumstances, although an increased yield of coffee results, the overall operation can be uneconomic. If no waste is available from a processing operation nearby and no grassland to hand which can be cut, mulch material must be grown on the plantation. This can require an area equal to that of the coffee. The mineral nutrient content of mulch material can affect the balance of nutrients in the soil over a period of several years. Elephant grass has a high potassium content, while sisal waste has a high calcium content. Either of these used continuously can affect the coffee adversely by altering the balance of nutrients (Mehlich, 1965, 1967). Mulch material should be changed at intervals to avoid this. Mulch materials are a fire risk in dry weather. It is usual practice to mulch alternate interlines each year. This should hinder the spread of fire and minimizes the cost of the mulch and its handling. Research in Kenya was reviewed by Njoroge (1989).

PRUNING

This is a key operation in coffee farming. Without pruning, the trees would become too large and the crop small and inaccessible. Mitchell (1974) showed that pruning had been found essential in all the important growing areas, and that a variety of methods were used. Andrade (1988d) confirmed the reasons why pruning is essential.

Pruning is essential for the following reasons.

1. To ensure that there is the maximum possible number of 2-year-old plagiotropic stems (primaries) on which the crop is borne.
2. To shape the trees to make the maximum use of the available space while not making access difficult.
3. To maintain an open environment within the tree so that light and air can reach all parts and the spread of pests and diseases is minimized.
4. To ensure that the crop is easily available for harvesting.

5. To remove diseased and over-age wood.
6. To enable chemical sprays, where used, to reach all parts of the tree.
7. To minimize biennial bearing and consequent risk of dieback.

There are two basic systems of pruning: single-stem and multiple-stem.

Single-stem Pruning

Formation

In order to stimulate branching from a single-stem tree, the main stem is cut back one or more times before it reaches its full height. To maintain the single stem to the full height of the tree, only one orthotropic shoot is allowed to grow from below the cut ('Colombia' pruning) (Andrade, 1988d).

Development

A coffee plant with a single orthotropic stem will produce plagiotropic branches (primaries) along the length of this stem. In their second year, these branches will flower along the length which grew in the first year. The next year, they will flower on the length which grew in the second year, while the part which has already flowered will be bare of leaves and flowers. The part of the branch carrying fruit will therefore become further from the main stem year by year. Some secondaries may grow from the second-year wood to fill the spaces between primaries. In due course, usually after 3–5 years, primaries die back. In arabica, they will remain attached to the stem if not cut off. In robusta, they usually drop off.

As a single-stem tree ages, the lower primaries die. The stem continues to grow and the section of stem carrying fruiting primaries moves upward. Stem internodes shorten as the tree grows taller. As a result, there often develops a mass of closely spaced primaries at the top of the stem and a bare stem below, so that the tree resembles an umbrella. Flowering is restricted in the mass of foliage and harvesting is difficult. The crop zone is often too high for the harvester to reach without a ladder and crop yields fall. Some cultivars, however, produce the highest yields in single-stem formation – for example, Catimor in Papua New Guinea (Kiara and Naged, 1995a).

Maintenance or rejuvenation pruning

The stem must be cut off at 40–50 cm from the ground. A number of new stems will grow from the stump remaining. If the foliage is well clear of the ground, stems may start to grow before the stem is cut. If the tree is to follow another single-stem cycle, all stems but one are removed. Alternatively, the tree can be changed to a multiple-stem system.

The height of a single-stem tree can be limited by capping at a height of about 2 m. The rate of increase of height can be restricted by capping at three intermediate levels, for example 0.2 m, 0.7 m and 1.2 m, with a final capping at 1.6–2.0 m. The capping height must be related to the height of the harvesters; short people cannot reach the crop on trees capped at 2.0 m. Each tree will need a light pruning annually, removing dead branches and surplus secondaries, in order to maintain the foliage open for maximum flowering. Careful control of foliage density will help to minimize biennial bearing. Viroux and Petithuguenin (1993) reported that capping was economically advantageous for robusta coffee.

Multiple-stem Pruning

Formation

In this system, from two to eight stems are allowed on each stump. Primaries branching outwards from the centre are encouraged, while those towards and crossing the centre are removed. The weight of primaries pulls the stems outwards, which opens up the centre of the tree. The original single stem may be cut in order to stimulate the production of several stems. This can be done in the nursery, but the plants will need to remain there longer to allow the new stems to grow and they will be bulkier to handle. Otherwise, the stem can be cut in the field when the tree is established, at 40–50 cm when the tree is about 70 cm high. Suckers will grow from below the cut; the strongest two should be selected for the first cycle and others removed. If two stems are permitted from each of two cuts a 'candelabra' tree is formed ('Costa Rica' pruning). Cutting the main stem more frequently (at heights of 0.75, 1.00, 1.25 and 1.50 m) produces a profusion of plagiotropic branches and a dense canopy. This technique is known as 'rock and roll' pruning (Andrade, 1988d).

Alternatively, the stem may be bent over and pegged down when it is about 50 cm tall. The lower primaries should first be removed. New stems will grow upwards from buds on the upper side of the stem. The strongest ones should be selected and the others removed. The stem must remain pegged down until it has set in its new position; it can then be cut back to just beyond the furthest new stem. This method is known as 'agobiado' and trees formed in this way as 'agobio' in South America. Planting trees at an angle of about 30° to the soil gives the same effect as bending. New stems will grow from the upper surface (see Fig. 4.2, p. 35). The trees are not so firmly anchored as in vertical planting, as the roots are not so deep initially. The roots are also to one side of the ultimate centre of the tree. These weaknesses can allow trees to be blown over by a strong wind. Multiple-stem trees can also be produced by the Brazilian 'Cova' system (p. 60). Maintenance of the field is difficult and expensive while the plants are small, but multiple-stem trees are formed quickly.

Development

Each stem produces primaries, as described for single-stem pruning. This system permits removal of individual stems when its branches have become too long or are at too great a height.

Maintenance or rejuvenation pruning

Stems have to be renewed at intervals, after 4–6 years in most cases. There are two approaches to this. One or perhaps two stems are removed each year. The number of stems retained should ensure that each stem has the life appropriate for local conditions. When new stems grow each year from the base of the tree, the appropriate number should be selected and the others removed. At the same time, all surplus primaries, particularly those crossing the centre of the tree, and also dead and diseased wood should be removed. Alternatively, all the stems can be cut off at the same time. This is known as 'stumping' the trees. The trees will then be completely renewed in the same way as newly planted trees. Often one or two stems only are left for 1 extra year; these are known as 'lungs' or 'breathers'. They are cut off the following year when the new stems are established. The lungs help the tree to grow new stems quickly. This system is illustrated in Fig. 6.8.

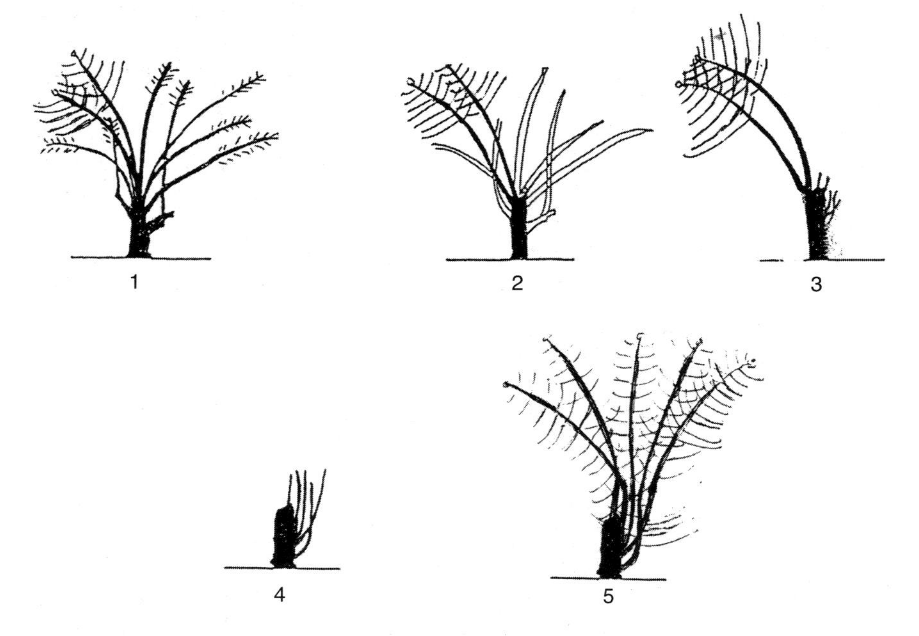

Fig. 6.8. Maintenance or rejuvenation pruning of multiple-stem coffee. 1, Old tree carrying unpruned stems; 2, old stems removed (shown in outline only); 3, new stems sprouting from stem now exposed to light; 4, 'lungs' removed after cropping; 5, new stems growing to re-form tree. (Drawn by Dr J.B.D. Robinson.)

Stems that are formed after pruning may be allowed to grow unchecked until they are removed at the next appropriate occasion of pruning; this is called 'tall multiple-stem pruning'. Alternatively, the new stems may be capped 1–2 years after pruning, and then allowed to continue growing until they are removed at the end of a 3-, 4- or 5-year pruning cycle. This system is 'capped multiple-stem pruning'. Plantings of dwarf cultivars are commonly managed by tall multiple-stem pruning with stumping. The stumping is sometimes carried out one block or field at a time, or otherwise by lines that are grouped according to the length of the cycle. Thus, using a 4-year cycle, for example, lines 1, 5, 9, etc., are pruned in year 1, lines 2, 6, 10, etc., in year 2, lines 3, 7, 11, etc., in year 3, lines 4, 8, 12, etc., in year 4 and the cycle repeated starting in year 5.

The choice of pruning system depends on local conditions and factors such as the cultivar and the spacing. For most coffee-growing areas, specific recommendations have been made.

IRRIGATION

This is an important part of the cultural system in several coffee-growing areas. It has, in many areas where rainfall is marginal, enabled coffee to be a profitable crop and has prevented total crop failure in dry years. Since flowering is induced by a change of internal water tension following rain after a dry season, irrigation can be applied to induce flowering.

Irrigation early in the dry season stimulates strong vegetative growth. This will provide plenty of nodes for fruit the following year. This can be used to ensure, as far as possible, that vegetative growth is even from year to year, which should minimize biennial bearing.

The amount of water required can be calculated from meteorological data or from evaporation measurement by a pan and the crop factor, as discussed in Chapter 5.

CHEMICAL MODIFICATION OF FRUIT ABSCISSION AND RIPENING

Overbearing could be minimized if part of the crop could be removed at an early stage. Trials have shown that Ethephon (2-chloroethane phosphoric acid) will promote the loss of up to 65% of the fruit. The same chemical accelerates the ripening of fruit. The effect is mainly on the pulp. In chemically ripened fruit, the bean has not developed fully. Gibberellic acid has also been found to affect fruit ripening. The result varies with concentration and time. Further investigation of these effects is justified by the potential for improvement, particularly for mechanical harvesting, by shortening and concentrating the ripening period.

HARVESTING

In many areas, coffee is harvested as the ripe cherry. The major proportion of this is done by hand. There is, however, a seasonal demand for labour and this can create problems. Harvesters (pickers) cannot be employed throughout the year and it is often difficult to find sufficient labour at the harvest season. Manual harvesting has one advantage in that the pickers can select only the ripe cherry, leaving the immature for a future occasion.

In Brazil, most arabica coffee is allowed to dry on the trees and picked as dried fruit. In Uganda, cherries are picked when red and sun-dried on mats; this product is known locally as 'Kiboko'. The same product, which is produced on a small scale in Kenya and Tanzania, is known there as 'Mbuni'.

Work on mechanical harvesters has been reported. Two types have been tried so far. The first, developed in Brazil after early experiments in Hawaii (Wang

Fig. 6.9. Front view of Brazilian coffee harvester, showing the oscillating fingers which shake the cherries off the tree (photograph from International Coffee Organization, London).

and Shellenberger, 1967), comprises rotating vertical cylinders carrying oscillating fingers, one each side of the row of trees. The vibration of the branches which is caused loosens the cherries, which fall on to a collecting conveyor (Fig. 6.9). This machine collects, at the appropriate frequency of oscillation, 96–98% of the cherries on the tree, of which 7–9% are immature or overripe (Watson, 1980). A machine incorporating vibrating fingers has been described. It is claimed to pick only mature cherries but it is not yet out of the developmental stage. Winston and Norris (1993) reviewed the development of mechanized systems in Australia. Synchronization of anthesis, which shortens the period over which ripe fruit can be harvested, reduces the amount of under- and overripe fruit harvested at one pass by machine (see Drennan and Menzel, 1994).

7

Mineral Nutrition and Fertilizers

Nutrients are removed from the plantation in the fruit, which comprise pulp and beans, and in prunings. The pulp is discarded; it is not normally weighed but the weight can be calculated from the weight of beans. The stems removed in pruning are usually carried out of the fields, thus removing nutrients. Estimation of nutrient loss from the field is therefore not straightforward, but several writers have reported calculated figures. From these data, it can be said that major nutrients lost for a production of 1 t of green beans are, approximately: 35–60 kg nitrogen (N), 6–12 kg P_2O_5 and 50–65 kg K_2O. These data indicate the order of need to maintain production and the relative amounts of the various nutrients.

Apart from applied fertilizer, there are inputs from the soil by root absorption and by decomposition of leaf fall, small prunings, shade-tree leaves and mulch. There are losses by leaching through the soil and by rainfall runoff from the soil surface. Ultimately, fertilizer needs and the effects of fertilizers on the crop yield have to be determined by experiment and plantation practice by economic factors.

Rao and Ramaiah (1986) reviewed all aspects of mineral fertilization of coffee. Accorsi and Haag (1959) reported the symptoms of deficiencies and excess levels of the macronutrients nitrogen, phosphorus (P), potassium (K), calcium, magnesium and sulphur. Loué (1957) reviewed the nutrition and fertilization of robusta coffee. Barel and Jacquet (1994) reviewed the effects of fertilizers on the quality of coffee.

EFFECTS OF INDIVIDUAL NUTRIENTS

Nitrogen

This element is essential for production of vegetative material. Where growth and production are low, under heavy shade, the nitrogen needed can be obtained from the soil. When growing conditions favour increased production, unshaded or under light shade, insufficient nitrogen will be available. Without fertilizers, a deficiency will arise and restrict production.

Under most conditions of commercial production, applying nitrogen will increase crop yield provided adequate supplies of other nutrients are available. In practice, levels of application from 50 to 400 kg N ha^{-1} are used. The amount depends on the local conditions and the level of yield expected. Low total applications are often made in a single annual application. Higher levels are always split into several applications. These are most effective when given at a time of high demand and low nutrient availability in soil during the wet season. Applications can be further divided into smaller parcels, but the times of application can have significant effects on crop yields. Local advice is available in most growing regions.

Several nitrogenous fertilizers are in common use. Sulphate of ammonia is widely used but acidifies the soil, which reduces the availability of base nutrients. Ammonium nitrate-based materials are less acidifying. Calcium ammonium nitrate is frequently used. Depending on its composition, it can be neutral or mildly acid. The calcium in the material may upset the balance of base nutrients if used continuously; this factor must be borne in mind. Urea has a high nitrogen content and is often used. Applied in unsuitable climatic conditions (hot, dry weather), much nitrogen can be lost. It contains an impurity, biuret, which affects the coffee if its concentration is too high.

Phosphorus

Only a small quantity of this element is removed with the crop and prunings. It is vital for plant growth, but may be in short supply due to fixation in the soil. Phosphate fertilizers are often applied, although direct responses to phosphate application to mature coffee are rarely seen in experiments.

The use of phosphate fertilizers in nurseries and on field planting more commonly gives significant responses. Foliar sprays of nutrient solutions including phosphate accelerate the development of young plants. M'Itungo and van der Vossen (1981) found that the growth rate of seedlings was improved by foliar NPK applications or incorporation of farmyard manure in the potting soil. Addition of phosphate fertilizer to the farmyard manure/soil mixture eliminated the need for foliar feeding. Foliar application to mature plants is more effective than ground application, but the expense of application to large areas often outweighs the return on the extra crop.

Mixing a phosphate fertilizer with the soil from a planting hole usually produces a significant improvement in the growth rate of young plants.

The phosphate fertilizers in most common use are the three superphosphates. Single superphosphate has the lowest content of P_2O_5 but does contain sulphur. Double and triple superphosphates contain more phosphate and also calcium. Particularly on the more acid soils, the calcium may have an adverse effect on the proportions of base nutrients. Normal field application levels may be as high as 200 kg P_2O_5 ha^{-1} per annum, but more usually the annual amount will be about 100 kg P_2O_5.

Mono- and diammonium phosphates contain nitrogen as well as phosphorus. They are soluble in water, so are used in foliar sprays and as a constituent of some NPK mixtures.

Potassium

The amount of potassium in the crop is at least as great as that of nitrogen. It is vital for growth in all parts of the plant but particularly important for development of the fruit. It is mobile within the plant; if the external supply is inadequate, the element will be withdrawn from leaves and transferred to the fruit. This causes an insufficiency in the leaves. In severe cases, dieback may occur.

Deep fertile soils can supply sufficient potassium, for many years at least. Mulch, particularly of grasses with a high potassium content, can help to maintain an adequate supply for the coffee. Under such conditions, significant responses to potash fertilizer application are rare. Where the soil and mulch provide insufficient potassium, it is vital to apply potash fertilizer, which will have a significant effect on crop yield. High levels are given to high-yielding coffee, up to 500 kg K_2O ha^{-1} per annum. Potassium and magnesium and also potassium and calcium (Furlani *et al.*, 1976) are antagonistic. A high level of one can restrict input of the other. Extra potash fertilizer will probably be necessary on soils with high calcium or magnesium contents.

Muriate of potash (potassium chloride) is the fertilizer used most widely to provide potassium. Sulphate of potash contains sulphur, which may be necessary in certain situations. Either can be incorporated into NPK compounds, a variety of which are often used for coffee. Gonzalez *et al.* (1977) investigated the effect of each of the sulphates and chlorides on the sulphur and chlorine contents of the leaves.

Calcium

A small quantity of this element is lost in the crop and prunings. Most soils contain sufficient calcium. Deficiency symptoms are known, but rarely seen. Excessive quantities of calcium will inhibit the uptake of potassium.

Lime or limestone will provide calcium when required. Where the soil is very acid, lime is often applied. This is useful, and acceptable provided the quantity is not large enough to restrict potassium uptake.

Magnesium

The loss of magnesium in crop and prunings is not large. Sufficient must remain within the plant as its presence is vital, since it is a constituent of chlorophyll as well as taking part in various growth processes. It is very mobile in the plant and is transferred when necessary from old to young leaves. Deficiency symptoms are therefore not uncommon on old leaves but this has no effect on yield unless the deficiency is very severe and affecting young leaves also. Hernandez (1962) reported increases in crop yield following application of magnesium sulphate on an acid soil.

Magnesium is usually applied as kieserite, which is an impure form of magnesium sulphate, dolomite, which is magnesium carbonate, or dolomitic limestone, which contains magnesium as well as calcium. Epsom salt, pure magnesium sulphate, is used in foliar sprays. Magnesium is often added to NPK fertilizers. Application rates of 50–100 kg MgO ha^{-1} per annum are usually adequate.

Sodium

A very small quantity of this element is found in normal leaves in plants on normal soils. Toxicity symptoms have been reported in coffee fertilized with sodium nitrate. Severe toxicity symptoms have been seen in robusta coffee planted close to the seashore.

Sulphur

This element is necessary in a small quantity. Some soils are deficient in sulphur and deficiency symptoms can appear in coffee on such soils. Many common fertilizers contain sulphur and it is common practice to include such a material, for example NPK, in a fertilizer programme. In some soils, sulphur deficiency symptoms will appear on young plants after planting out. To prevent this, a sulphur compound is mixed in the soil before planting. Gypsum, calcium sulphate, is often used. Sulphate of ammonia should not be put into planting holes. Sumbak (1983) reviewed reports of sulphur deficiencies, remedial fertilizer applications and the sulphur content of plants.

Zinc

The characteristic deficiency of zinc is often seen. If severe, there will be a significant reduction of yield. It is easily remedied by a foliar spray of a zinc compound, either zinc sulphate solution or a suspension of zinc oxide or zinc oxychloride. Ground applications are not very effective. Ananth *et al.* (1965) described zinc deficiency symptoms, found no symptoms of toxicity and confirmed the efficacy of foliar applications of zinc compounds in curing the deficiency.

Copper

A deficiency in coffee in the field has never been reported. The regular use of copper fungicides maintains leaf content above the deficiency level. Symptoms of copper toxicity due to high levels in the leaf have been described, but it is doubtful if these occur in the field. Maroko (1991) reported levels of copper in soils and plants where copper-based fungicides were applied twice yearly.

The continuous use of copper fungicides has been found to increase crop yield. It has not been possible to relate this to the copper content of the leaf, but this 'tonic' effect is well authenticated.

Iron

A deficiency of iron occurs frequently on the more alkaline soils or soils with a high organic content and is often seasonal in drought and not serious. It is often associated with high phosphorus or manganese levels. Acidification of soil with sulphur or sulphate of ammonia helps to minimize iron deficiency. Application of an iron chelate will cure the deficiency. Insertion of a solid iron salt is also effective but a foliar spray of a salt is not. Robinson (1960) showed that iron deficiency affected the quality of coffee beans, causing 'amber beans'.

Manganese

Deficiencies occur occasionally on alkaline soils. Toxicity symptoms, known as 'cafe macho' in Costa Rica, are found on acid soils. Correction of the soil acidity level usually cures the problem.

Aluminium

This element is freely available in acid soils. It is believed that it is not essential for coffee. Toxicity effects have been reported for arabica and the critical level in leaf above which symptoms occur has been defined (Pavan *et al.*, 1982).

Boron

Both deficiency and toxicity symptoms are known for this element. The deficiency is often associated with high calcium in the soil, but a low boron level in the plant restricts calcium uptake. The ratio of boron to calcium is important. High application levels of potassium may affect the boron content by restricting calcium uptake. The deficiency can be cured by an annual application of 30–60 g borax to each tree or by a foliar spray of a 0.4% solution of borax.

Molybdenum

A deficiency of molybdenum has been described (Malavolta *et al.*, 1961) and a minimum critical level established. The deficiency may arise on very acid soils.

Chlorine

Toxicity symptoms have been described and a critical leaf content established. Application of muriate of potash was found to raise the chlorine content to a level close to the critical value. Severe damage has occurred on robusta coffee close to the seashore. Gouny (1973) investigated the response of coffee to chloride ions.

ORGANIC MANURE

Organic manures are valuable and can be used in several ways. Incorporation in nursery soil stimulates the growth of young plants, while the common practice of incorporation in planting holes stimulates growth. Put into the soil around a tree planted in the field, manure accelerates the growth of young plants. Spread as a mulch under mature coffee, the soil is improved and protected against erosion, while nutrients are released to the soil as the mulch breaks down. Almost any organic material that is conveniently to hand can be used. The handling of bulky material is expensive.

The effect of organic material on the ratio of bases in the coffee must be kept in mind. Some materials which are high in one base can, with regular use over a period of years, raise the amount of the base in the soil to a level at which a toxicity effect will arise or the uptake of other nutrients be restricted (Mehlich, 1965).

NUTRITION AND CROP QUALITY

The relationship between nutrition and quality has been investigated on several occasions and reviewed by Mehlich (1967). Nitrogen reduces quality, as measured by size of beans. The reduction is more severe on soil of low fertility when the levels of other nutrients are inadequate. Quality is reduced when the balances between the various bases move away from the optimum. Iron deficiency on soil with a high pH value produces 'amber beans' (Robinson, 1960).

FOLIAR ANALYSIS

Chemical analysis of selected leaves has been used successfully in many crops to monitor nutrition and as a basis for establishing fertilizer programmes. For this approach to be successful, the leaves sampled must always be the same in terms of age, location and condition. In practice, the third and fourth leaf pairs on an active primary branch bearing crop give results that are reproducible to the degree where the data can be used for diagnosis of abnormalities and determination of fertilizer requirements. Table 7.1 lists data used successfully by the author (Willson, 1985). Each sample should consist of 100 leaves selected in a random manner from trees in the block being sampled. Snoeck (1984a) reviewed the use of plant analysis to determine nutrient requirements. Harding (1993) investigated the seasonal fluctuations in leaf nutrient contents and related these to rainfall and the crop development cycle. Interpretation of foliar analysis data must be related to the stage in the growth cycle at which the samples are taken; accurate interpretation depends on consistent timing of reference data and sampling.

Koffi (1995) determined the optimum level of available potassium and the optimum ratio of magnesium to potassium for robusta coffee in Côte d'Ivoire.

SYMPTOMS OF NUTRIENT DEFICIENCY AND TOXICITY

- *Nitrogen deficiency*. The older leaves lighten in colour, becoming yellow and sometimes almost white. Parts of leaves in direct sunlight scorch and fade more quickly. The leaves then fall and stems die back from the tips.
- *Phosphorus deficiency*. The leaves become mottled with yellow spots showing a red tint. If the deficiency is severe, the yellow chlorosis covers the whole of the leaves, which then fall off.
- *Potassium deficiency*. The older leaves scorch at their tips and margins. Most of the affected leaves fall off, leaving branches carrying only a few leaves. These branches later die back.
- *Calcium deficiency*. Young leaves become irregularly chlorotic (yellowish)

Table 7.1. Critical levels of nutrients in leaves of coffee (leaves from third or fourth pair on a shoot). (From author's values, based on reconciliation of published data.)

	N (%)	P (%)	K (%)	Ca (%)	Mg (%)	S (%)	Cl (%)	Fe (p.p.m.)	Mn (p.p.m.)	Zn (p.p.m.)	Cu (p.p.m.)	B (p.p.m.)	Al (p.p.m.)	Mo (p.p.m.)
Arabica														
Deficient														
	2.00	0.10	1.50	0.40	0.10	0.10		40	25	10	3	25		0.5
Subnormal														
	2.60	0.15	2.10	0.75	0.25	0.15		70	50	15	7	40		0.8
Normal														
	3.50	0.20	2.60	1.50	0.40	0.25	0.2*	200	100	30	20	90	60	
Excess														
Robusta														
Deficient														
	1.80	0.10	1.20	0.40	0.20	0.12		40	20	10	13	20		0.3
Subnormal														
	2.70	0.13	1.80	0.80	0.30	0.18		70	35	15	20	35		0.5
Normal														
	3.30	0.15	2.20	1.50	0.36	0.26		200	70	30	40	90		
Excess														

N, nitrogen; P, phosphorus; K, potassium; Ca, calcium, Mg, magnesium; S, sulphur; Cl, chlorine; Fe, iron; Mn, manganese; Zn, zinc; Cu, copper; B, boron; Al, aluminium; Mo, molybdenum.
*Tentative value.

on the margins, leaving an area with saw-like edges around the main veins, and become convex. In severe cases the older leaves become chlorotic at the tip and cork is formed on the larger veins on the underside. Sometimes the axillary bud dies.

- *Magnesium deficiency*. An orange-yellow to brown chlorosis starts near the centre vein and spreads between the veins. A narrow area each side of the main and subsidiary veins remains green. This deficiency always shows initially on the older leaves, which will fall off.
- *Sodium toxicity*. This is visually the same as potassium deficiency.
- *Sulphur deficiency*. The younger leaves change to a yellowish-green colour. This chlorosis starts as a band around the main vein. The lower surface of each leaf is lighter than the upper. Some mottling occurs near the leaf margins.
- *Iron deficiency*. The younger leaves become pale green and later yellowish and almost white. All the veins retain their green colour until the deficiency becomes very severe.
- *Manganese deficiency*. The younger leaves become uniformly pale olive-green with small yellow spots. The older leaves in direct sunlight become lemon-yellow in colour.
- *Manganese toxicity*. Root development is restricted and leaves become small, deformed and orange in colour. Tree vigour and flowering are reduced.
- *Zinc deficiency*. The leaf veins form a green network against a pale green to yellow background. In severe cases, the leaves become almost white, spear-shaped and brittle. Very characteristic is a drastic shortening of the stem internodes and multiple bud growth (rosetting) at the nodes. Branches may eventually die back.
- *Copper deficiency*. The younger leaves distort to an S shape and lose their green colour before developing necrotic patches on the margins.
- *Copper toxicity*. All leaves lose their colour slightly and fall off.
- *Boron deficiency*. The leaves become smaller, olive-green at the distal end, narrow and twisted, with irregular edges and a rough surface. The terminal bud dies and branches grow out like a fan.
- *Boron toxicity*. All the leaves lose their colour and necrotic spots develop on them.
- *Aluminium toxicity*. The margins of younger leaves become necrotic and leaves curl because the centres grow faster than the margins.
- *Molybdenum deficiency*. Yellow spots develop near leaf margins and become necrotic, those in the centre first. The leaf blades curl so that the sides touch each other underneath the leaf centre.

Coffee can also suffer from two other leaf abnormalities, as follows.

- *Biuret toxicity*. Biuret is an impurity in urea. The marginal interveinal regions of the leaves lose their colour and the leaves curl upwards ('cupping'). Chlorosis can also develop in irregular spots on the leaves.

- *Hot-and-cold disease*. See p. 97 in Chapter 8.

The Coffee Board of Kenya (CBK) has published descriptions of the more important deficiency conditions, with coloured photographs (CBK, 1970c).

8

PESTS, DISEASES AND WEED CONTROL

Both pests and diseases can cause major problems in coffee, leading to losses of crop and to trouble and expense for growers. Fortunately, the final product, coffee beans, suffers little from either pests or diseases during storage if it is properly dried and stored in good conditions where it will keep dry. Arabica coffee is much more susceptible to pests and diseases than robusta coffee. The major part of the text which follows refers specifically to arabica; the extent of the problems affecting robusta is discussed at the end of each section. Le Pelley (1968) dealt comprehensively with coffee pests, with many references. The Coffee Board of Kenya (CBK) has published descriptions of pests (CBK, 1970a) and diseases (CBK, 1970b) affecting arabica coffee in that country, with coloured photographs of some and recommended treatments. Andrade (1988e) lists South American pests and diseases with descriptions, coloured photographs of some and control measures. Lavabre (1961) described many pests in detail, including some which attack robusta coffee, and gave brief notes on some diseases. Many of the recommended treatments are now out of date. Andrade (1988c) lists the weeds of coffee in Venezuela and discusses control. The effects of these factors on quality have been discussed by Barel and Jacquet (1994).

PESTS

Cramer (1967a) estimated that the losses of the potential coffee crop from pest attack were: Africa 15%, Asia 10%, America 12% and Oceania 8.3%. The greatest number of pests occur in Africa, as might be expected from coffee's African origin. Latin America has relatively few, but losses of coffee are substantial.

Over 900 species of insect have been reported as feeding on coffee but relatively few create sufficient damage and are sufficiently widespread to become major problems.

Methods for pest, disease and weed control must be viewed as integral parts of the management of a coffee plantation. Cultural, chemical and harvesting methods must be planned as an entity. Integrated pest management has been very successful in East Africa.

Pests of Foliage

Leaf miners (Fig. 8.1)

Major damage is caused by *Leucoptera meyricki* in Africa, *Leucoptera coffeella*, *Leucoptera meyricki* and *Leucoptera caffeina* in Central and South America. Damaged leaves have brown irregular blotches on the upper side; if broken across the blotch by bending, the mine is opened, exposing white caterpillars within. The leaves die and fall prematurely. Major defoliation can result, causing severe loss of crop. Routine application of insecticides to the trees often results in more severe damage, as natural enemies are reduced in number. Applications must be related to moth counts and examination of damaged leaves. Control has also been established by spreading a granular systemic insecticide to the soil. This controls insects attacking stems and foliage but does not harm beneficial insects. The CBK (1970a) recommends fenitrothion, fenthion, parathion or dicrotophos. Eight insecticides were found to be effective in South America (Andrade, 1988e).

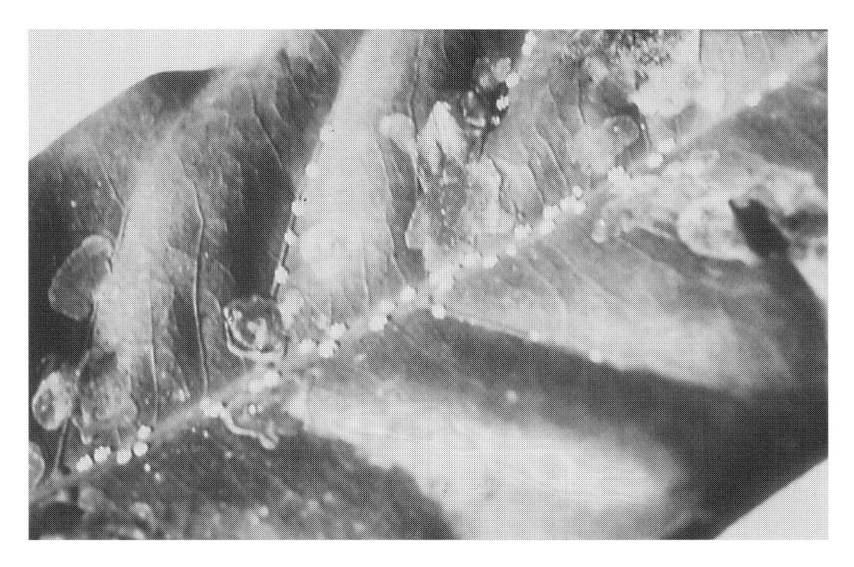

Fig. 8.1. Coffee stem and leaves infested by leaf miner (*Leucoptera* spp.) and star scale (*Asterolecanium coffeae*) (photograph from International Institute of Biological Control).

Caterpillars

The giant looper, *Ascotis selenaria reciprocara*, has occasionally become a serious pest in East Africa. These outbreaks have been attributed to adverse effects on natural enemies arising from excessive use of organophosphorous insecticides. *Bacillus thuringiensis* selectively destroys *Ascotis* (Waikwa and Mathenge, 1977). Several other types of caterpillars attack leaves to varying degrees; the green looper (*Epigynopterix stictigramma*) is a minor pest in Kenya (CBK, 1970a). The CBK recommends methomyl.

Mealybugs

In Kenya, the common coffee mealybug *Planococcus kenyae* devastated coffee about 60 years ago. White masses of insects form on stems, honeydew or sooty mould on leaves. The root mealybug *Planococcus citri* can be troublesome in poor soil; roots become covered in white fungus (*Polyporus* spp.) covering the pests. Control of the former in Kenya has been achieved by classical introduction of the parasite *Anagyrus kivuensis*. Ants feed on honeydew produced by mealybugs and interfere with natural enemy activity. Control of the ants is important. The CBK (1970a) recommends diazinon to control the common mealybug. Dursban is also recommended for controlling ants.

Scale insects (Figs 8.1 and 8.2)

Green scales (*Coccus* spp.), brown scale (*Saissetia coffeae*), white waxy scale (*Ceroplastes brevicauda*) and star scale (*Asterolecanium coffeae*) are all found on

Fig. 8.2. Coffee branch with infestation of green scales (*Coccus* spp.) (photograph from J.M. Waller, CABI Bioscience).

coffee in East Africa. *Coccus* spp. are known in India and have been accidentally introduced elsewhere, e.g. Papua New Guinea, while *Aspidiotus* spp. and *Icerya pattersoni* have recently achieved pest status. They can be controlled by pruning heavily infested tress and applying nitrogen fertilizer and mulch. Painting white oil or tar oil on infested bark destroys the insects. The CBK (1970a) also suggests azinphos-methyl or malathion. Andrade (1988e) lists furadan, temik, lebaycid, nuracron and basudin. Samuel *et al.* (1993) reported an indirect control method, using insecticide and ants.

Cultural practices are important; for example, branches that touch the ground must be removed to eliminate alternative access when stems are banded with insecticide. For light infestations of white waxy scale, branches can be pruned off and left on the ground; natural enemies will emerge and clear the scales. Barriers of dust prevent access of star scale.

Mites

The yellow tea mite, *Hemitarsonemus latus*, and red spider mite, *Olygonichus coffeae* occur sporadically. Predators usually clear infestations quickly. Andrade (1988e) lists dicofol, dimethoate and phosphamidon for control of red spider mite. Control of red spider mite by spraying a suspension of sulphur is sometimes necessary (CBK, 1970a).

Other insects

Foliage is attacked to varying degrees by a number of other insects. Thrips (*Thysanoptera* spp. and *Diarthrothrips coffeae*), lacewing bugs (*Habrochila* spp.) and mirid bugs (*Miridae*) are all seen at times. The mirid *Stethocomus* spp. feeds on lacewing nymphs and adults. All these species have natural enemies, and chemicals should only be used when the predator population is low (CBK, 1970a).

Epicampoptera spp. (tailed caterpillars or chenilles queue du rat) are normally minor pests of arabica coffee but occasional severe attacks can defoliate trees, followed by eating of fruit and green bark. The caterpillars can grow up to 6 cm long; they have many predators but severe attacks need chemical control (Lavabre, 1961; CBK, 1970a).

Pests of Stems

Borers (Fig. 8.3)

The yellow-headed borer (*Dirphya nigricornis*), the white borer (*Anthores leuconotus*), the West African coffee borer (*Bixadus sierricola*), *Apate monachus*, *Apate indistinctus* and other *Apate* spp. and the black borers all damage stems in Africa. *A. leuconotus* larvae girdle the bark at the base of the stem and then bore into the tap root. They then travel upwards inside the stem, emerging 8 months later. *Dirphya nigricornis* eggs are laid on primary branches. The larvae

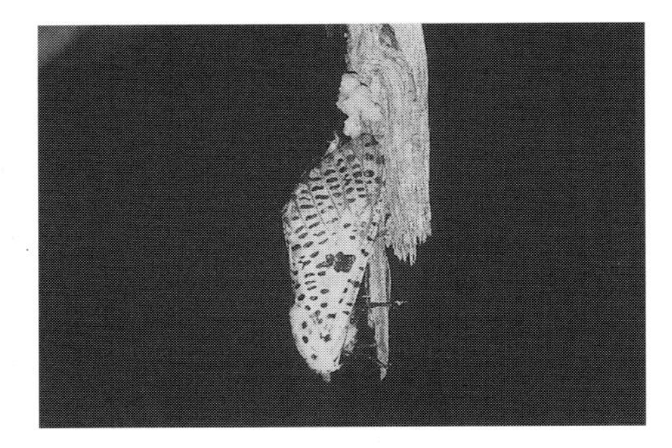

Fig. 8.3. Coffee stem borer (*Xylotrechus quadripes*) adult moth (photograph from International Institute of Biological Control).

bore downwards inside branch and stem, ejecting grass from holes periodically. Eventually they exit through a larger hole. White stem borer (*Xylotrechus quadripes*) is the most serious pest in India. *Xyleborus* spp. are also troublesome. Throughout Africa, a number of moth larvae, for example *Eucosoma nereidopa*, bore into green shoots and work downwards. The shoot tip wilts and dies; breaking off the tip will often remove the pest also. Stems are weakened and can be broken easily so that the crop on that stem is lost. Parasitic wasps attack the larvae; control can be achieved by removal of wilted primaries and spearing larvae in their holes (CBK, 1970a).

Pests of Flowers, Fruit and Beans

Antestia bugs

The variegated coffee bug (*Antestiopsis orbitalis*) can cause severe damage to the crop in Africa. They suck young berries, causing longitudinal, brown, zebra stripes, which are obvious on washed beans, and occasional empty beans. The striped beans, zebra beans, produce a foul odour, which spoils the product if they are not removed. The insects feed on stem terminal buds in the absence of fruits, so that a fan of stem growth is formed. They also introduce a fungus (*Nematospora* spp.), which makes beans rot to a white paste; such beans in processed beans are called 'posho beans'. While empty beans float and are separated in the washing, zebra beans are not separated. Control of tree habit by pruning to keep the canopy open minimizes attack by antestia; the bugs prefer a dark, enclosed habitat. Sample trees are sprayed monthly with insecticides and the number of dead bugs counted. If, on average, more than two are found per tree in dry areas or one per tree in wet areas, all trees should be sprayed with pyrethrum or an organophosphorous insecticide to kill the bugs. These insects are parasitized by wasps (CBK, 1970a).

Mirid (capsid) bugs

Lamprocapsidea (*Lygus*) *coffeae* attack and kill flower buds. The test spraying for antestia also kills the capsids. If there are ten or more per tree, all trees must be sprayed. A very low rate of insecticide application controls antestia and capsid, which minimizes the effect on harmless and desirable insects (CBK, 1970a).

Berry borer (Fig. 8.4)

Hypothenemus hampei, previously known as *Stephanoderes coffeae*, originated in Africa but has spread to America and Asia. The females bore into the beans on the tree, where the larvae feed inside. After harvest, the beetles breed in dry or overripe beans on the ground. Many damaged beans fall from the tree; those that do not fall are defective and reduce crop quality. This pest occurs in Africa (CBK, 1970a), Asia and South and Central America, except Venezuela (Andrade, 1988e). Heavy shade discourages the parasites, while frequent harvesting minimizes the number of infested beans. Castro (1990) reported recent research in Central America, including the variations in susceptibility of various cultivars, control by manual removal of berries and chemical control, using endosulphan. Galvez (1992) discussed the introduction of African parasites to Central America to give biological control of this pest. Montagnon and Leroy (1993) investigated the resistance of clones of *Coffea canephora* to berry borer in West Africa. Mbondji (1995) discussed biological control and reported that diazinon was particularly effective against *H. hampei*. Bustillo (1995) reported integrated pest management, including the release of biocontrol agents.

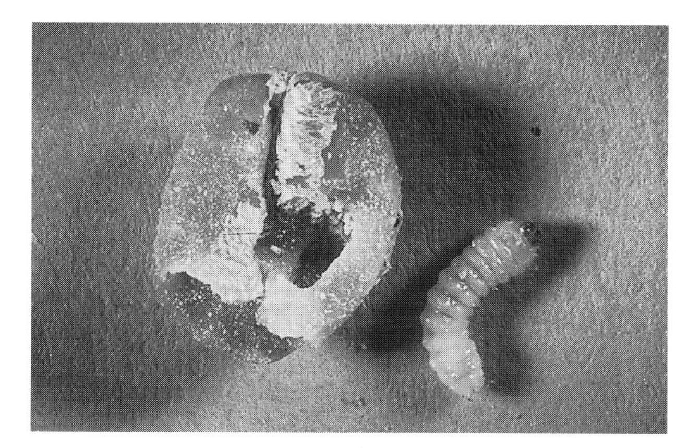

Fig. 8.4. Larva of coffee berry borer (*Hypothenemus hampei*) and berry damaged by this borer (photograph from J.M. Waller, CABI Bioscience).

Berry moth

Prophantis (*Thliptoceras*) *smaragdina* larvae bore into beans while they are developing. Three species of parasitic wasp have been noted in Kenya. Malathion can be sprayed before larvae enter beans (CBK, 1970a).

Mediterranean fruit fly

Ceratitis capitata lays eggs in the developing fruit, and the larvae feed on the mucilage. Some of the fruit fall but others remain on the tree. At one time, it was thought that this pest did not harm the product but it was found that the embryo in the seed was killed. This became infected and caused an undesirable flavour in the product. Known as 'hidden stinkers', these unpleasant beans could not be detected until many months later. It has since been established that infected beans fluoresce under ultraviolet light; colour-sorting machines have been installed to separate the infected beans. This pest has been a problem in East Africa, but no control is recommended (CBK, 1970a).

Beetles

Systates pollinosus and other *Systates* spp. feed at night on the edges of leaves, retiring to the soil by day. The dusty brown beetle *Gonocephalum simplex* feeds on bark and stalks of green cherries by night; by day they are in the soil. The larvae are known as 'false wireworms' and they eat seeds and attack coffee roots (CBK, 1970a).

Lesser coffee bean borer

Araecerus fasciculatus damages stored coffee ready for shipment. It transfers from the residues of other crops. Infected beans are fumigated to destroy these beetles.

Pests of Roots

Nematodes (Fig. 8.5)

Meloidogyne africana and other root-knot nematodes are widespread in East Africa, but the level of crop loss has not been established. *Meloidogyne caffeicola* and the lesion nematode *Pratylenchus coffeae* cause severe damage in some parts of India and Central and South America. *Radopholus similis* is an important pest in Asia. Soil application of granular systemic insecticides, such as carbofuran and phenamiphos, minimizes nematode damage and helps control scales, mealybugs and leaf miners. Andrade (1988e) reports favourable results from furadan, nemacur, temik and vydate. They are more effective on young and nursery plants. Problems in nurseries can be avoided by using fresh soil, never used for coffee, for each new nursery. Soil sterilization will also kill other insects, fungi and weed seeds. Kumar and Samuel (1990) reviewed the effects and treatment of all nematodes attacking coffee. They recommended planting infected areas with robusta coffee or plants grafted on to arabica–

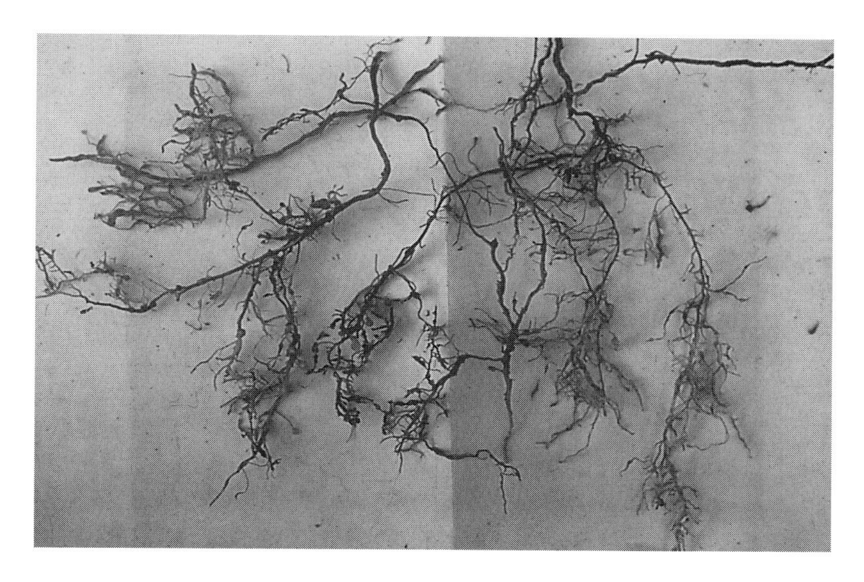

Fig. 8.5. Coffee roots damaged by root-knot nematode (*Meloidogyne africana*) (photograph from J.M. Waller, CABI Bioscience).

robusta hybrid rootstock. Campos *et al.* (1990a) described all nematodes affecting coffee and their effects on the trees. Eskes *et al.* (1995) reported an investigation into the susceptibility of a range of cultivars to nematode attack and proposed strategies for breeding resistant rootstock varieties.

Other Insects

Beetle larvae (chafer grubs) and moth larvae (cutworms) can damage roots. Cicada nymphs attack roots and are a persistent problem in Brazil. Crickets and snails attack stems at ground level. A persistent insecticide can be incorporated into nursery soil. Painting stems of young coffee in the field with a persistent insecticide protects the young plants.

Other Pests

Andrade (1988e) lists eight groups of minor pests with symptoms of attack and control.

Other Animal Pests

A number of birds and monkeys eat coffee berries, which contain the sweet mucilage. Monkeys cause damage by breaking branches. Trees can also be damaged by animals eating stem tips or rubbing against the trees. Molerats will damage roots.

INSECTS ATTACKING ROBUSTA COFFEE

Pests of Nurseries

Leaves are eaten by *Leucoplema* spp. (grey caterpillars). Stems are attacked by *Grillotalpa africana* (crickets) and cutworms (various species).

Pests of Leaves

Leaves are holed by *Melanotha* spp. (cockchafers). Leaves are eaten by *Epicampoptera* spp. (tailed caterpillars), *Cephonodes* spp. (giant loopers), *Zonocerus* spp. (locusts and grasshoppers) and *Dulinus unicolor*. Leaves are distorted by *Dichocrosis* and *Prophantis* spp. (coffee berry moths), *Pseudococcus* and *Coccus* spp. (scales and mealybugs) and *Toxoptera* spp. (thrips and aphids). Leaves are mined by *Leucoptera* spp. (leaf miners). *Epicampoptera* spp. (tailed caterpillars) are widespread in Africa.

Pests of Stems

Xylosandrus spp. (coffee twig borer) bore into twigs. Stem tips are bored by *Xyleutes* spp., while *Bixadus* spp. and *Anthores* spp. (stem borers) bore into the bases of stems. *Xylosandrus* spp. can be defeated by planting tolerant clones. *Xyleutes* spp. can be controlled by pruning and burning infected wood. All the others listed need insecticide treatment. Rats cut and damage young stems.

Pests of Berries

Hypothenemus hampei (coffee berry borer), *Dichocrosis* spp. and *Prophantis* spp. (coffee berry moths) bore into berries.

DISEASES

Diseases can cause severe losses of crop and crop potential, both directly, by destroying or damaging flowers and fruit, and indirectly, by reducing the foliage available for photosynthesis. Cramer (1967a) estimated that the losses from diseases amounted to: Africa 20%, Asia 15%, America 15% and Oceania 16.7% of the potential production. Waller (1988) reported the current position on coffee diseases.

Diseases of Leaves

Leaf rust (Fig. 8.6)

Hemileia vastatrix is the most important pathogen of coffee and is now present in all coffee-growing countries. It was eradicated in Papua New Guinea in 1965 but reappeared in 1985. Appearing initially in Sri Lanka in 1880, it destroyed the coffee industry in that country. The former coffee plantations changed to growing tea. It attacks arabica coffee much more severely than robusta, so in some countries, notably Indonesia, robusta has replaced arabica as the major crop. Hybrids between *Coffea arabica* and *C. canephora* are very resistant. Back-crosses using such hybrids, e.g. Catimor, are now being released. Montagnon and Leroy (1993) investigated the resistance of *C. canephora* clones.

The fungus produces orange powdery spots on the undersides of the leaves – hence the name of the disease, leaf rust. Diseased leaves fall prematurely. The next season's vegetative growth is therefore reduced, restricting the crop potential for the following year. Overbearing may result so that stems die

Fig. 8.6. Coffee rust (*Hemileia vastatrix*): fungal growths on undersides of leaves (photograph from J.M. Waller, CABI Bioscience).

back, leading to further reduction in yield and possible death of the tree.

Spores of the fungus are spread by wind and rain and also on persons and machines moving in the plantations. Infection needs rain so that it occurs only in the wet season. A long incubation time, 2–6 weeks according to temperature, results in major outbreaks occurring early in the dry season. At high altitudes, over 1700 m, cool weather prevents serious epidemics arising. The effects of the disease can be minimized by cultural measures. Trees should be pruned so that there is free circulation of air within the canopy and leaves can dry rapidly after wetting. Soil should be kept free of weeds and moisture conserved by mulching. Trees must not be allowed to overbear, and nitrogen must be applied to encourage the growth of new leaves. Holguin *et al.* (1993) studied the virulence and aggressiveness of various races of *H. vastatrix*.

The traditional treatment is the application of copper fungicides; in Kenya, 3–5 kg ha^{-1} and, in Venezuela, 3–7 kg ha^{-1} are applied at intervals of 4–6 weeks, starting 10–15 days after the rains start (CBK, 1970b; Andrade, 1988e). Systemic fungicides have been used, e.g. Bayleton at 1–2 kg ha^{-1} (Andrade, 1988e). He states that chemical control is not economic in low-yielding coffee (annual yield less than 920 kg ha^{-1}). The Kenya Coffee Research Station (1993) published revised recommendations for control, including techniques and application rates suitable for smallholders. Silva-Acuna *et al.* (1993) described control of leaf rust under shade by triadimenol combined with copper oxychloride in Venezuela. Triadimenol was applied alone to the soil and mixed with copper for application to the leaves. Alarcon and Carrion (1994) described the use of *Verticillium locanii* as a biological control of coffee rust.

Virulent strains have overcome resistance in some arabica cultivars, but hybrids with robusta are resistant. Resistance has been retained in some progeny of these hybrids from back-crossing of a hybrid with arabica. Epidemiology, resistance and management were discussed by Kushalappa and Eskes (1989). Agnihothrudu (1992) reviewed all aspects of this disease.

Brown eye spot

Cercospora coffeicola is a very common fungus causing disease. In South America, it can lead to sufficient leaf damage to defoliate trees and it affects the beans, causing 'parapara beans', which are difficult to pulp. Systemic fungicides, e.g. difolatan, antracol, dithane or benlate, are applied every 10–20 days to control this disease (Andrade, 1988e). Hanumantha *et al.* (1994) reported the effects of captafol and carbendazole on this disease. The former was more effective. Brown eye spot rarely causes sufficient damage in Kenya to justify control measures, although it can affect young plants in nurseries to a degree which justifies chemical control. It may be necessary to spray a copper fungicide in nurseries (CBK, 1970b). This disease prefers warm, humid conditions. Circular brown spots have a greyish centre and, sometimes, a yellow halo. They rarely grow larger than 5 mm diameter.

South American leaf spot

The fungus *Mycena citricola* occurs only in Latin America, where it can cause appreciable damage in cool, wet conditions. The buff to white circular lesions grow to 15 mm diameter and can appear on stems and berries as well as leaves. Heavy infection causes defoliation. Outbreaks are often localized, and control must start with thinning of tree canopies and shade to allow more light and air to reach the coffee leaves, together with removal of infected material and alternative host plants. Chemical treatment is sometimes necessary; lead arsenate was widely used, but copper compounds, particularly Bordeaux mixture (Andrade, 1988e) and tridemorph, are also effective.

Pink disease

Corticium koleroga, *Corticium salmonicolor* and other *Corticium* species infect bearing branches and destroy the crop thereon. It occurs mainly in coffee at low altitudes – below 1000 metres in Venezuela, where temperature and humidity are high. Pink mycelium spreads over the surfaces of stems, leaves and berries and kills individual branches. Removal of infected parts of the tree is often sufficient to control it; application of a copper fungicide – trimiltox or benlate – is recommended in Venezuela (Andrade, 1988e).

Phoma spp. and *Colletotrichum* spp.

These fungi can produce irregular, dark, necrotic lesions on old or damaged leaves.

Derrite

Phyllostricta caffeicola affects coffee at high altitudes in South and Central America, particularly in humid conditions and when affected by cold winds. The terminal leaves on stems are most severely affected. Lesions with fruiting bodies grow on terminal and adjacent leaves, which then distort and shrivel. Application of difolatan, captan or benlate controls this disease (Andrade, 1988e). Control by pruning and fungicide application was discussed by Chereguino (1979).

Bacterial blight (Fig. 8.7)

This is caused by the bacterium *Pseudomonas syringae* and produces dark necrotic patches on leaves, which spread to the young shoots and cause die-back. Copper fungicides are bactericidal and control this disease, but organic fungicides are not bactericidal. This disease became a problem when organic fungicides replaced copper. Removal of affected trees is recommended in Venezuela (Andrade, 1988e).

Fig. 8.7. Young coffee stems killed by bacterial blight (*Pseudomonas syringae*) (photograph from J.M. Waller, CABI Bioscience).

Diseases of Stems

Fusarium bark disease, Storey's bark disease, collar rot

Fusarium stilboides is important in Central Africa and is spreading in Kenya. It is widely distributed but attacks only susceptible varieties grown in warm, dry conditions. Scales form on the bark and develop into cankers. Green vegetative shoots are killed; this is known as Storey's bark disease. Affected trees die back. The most severe effects are seen at the collar (collar rot). Symptoms can be minimized by ensuring optimum growing conditions and applying fungicidal paint to infected bark. Frequent spraying of captan will control the disease, but this is very expensive (CBK, 1970b).

Tracheomycosis or vascular wilt disease

Fusarium xylarioides occurred sporadically in Africa but has recently turned virulent (Flood, 1997). A spiral yellowing of leaves develops; the yellowing spreads later to all leaves and the stem dies. Diseased trees must be removed from the plantation and the soil disinfected before another tree can be planted.

Elgon dieback

This disease is caused by the same fungus (*Colletotrichum kahawae*) that causes coffee berry disease (CBD). It occurs most frequently on Mount Elgon and occasionally elsewhere in Kenya. Nodes blacken and the blackening sometimes spreads to the adjacent leaves. The stem above dies, the leaves turning brown. The disease is minimized but not always controlled by copper sprays. Captafol increases the severity of attacks; benomyl can be used as an alternative to copper (CBK, 1970b).

Ceratocystis fimbriata

This fungus causes coffee wilt or canker in Latin America. Entering through wounds in the bark, the fungus causes dark, sunken, necrotic cankers. Trees die when the stem is girdled (Chereguino, n.d.). To control this disease, shade and coffee trees must be pruned to reduce the humidity around them. Infected parts are cut off. Almost 90% control has been achieved by painting wounds with a paste containing copper oxychloride.

Diseases of Roots

Govindarajan (1988) reviewed root diseases and their management.

Armillaria root rot

Armillaria mellea occurs frequently on coffee, usually transferring from the roots of former forest or shade trees. There is a rapid decline of affected trees. Creamy white mycelial strands are found under the dark and pale brown mushroom-like sporophores grown on the bases of recently killed trees. Slow killing of forest and unwanted shade trees by frilling exhausts the reserves of carbohydrates in the roots and the fungus dies out, minimizing the risk of transfer to the coffee. Old roots must be thoroughly removed. Affected coffee plants must be completely removed and burnt; the soil at the site of an infected tree should be dug out and left exposed for a period to kill any remaining *Armillaria*.

Black root rot

Rosellinia spp. attack coffee in Latin America, particularly in recently cleared land. Symptoms are similar to those of *Armillaria*, although the sporophores are minute spherical structures. Prevention and treatment are the same as for *Armillaria*, to which Andrade (1988e) adds applying a 1% solution of Brasicol to the roots of coffee.

Fusarium solani

This serious fungal root disease occurs sporadically in Africa, particularly in dry areas. Prevention and treatment are similar to those for *Armillaria*.

Lyamungu dieback

This is a physiological condition, which occurs mainly in northern Tanzania. It affects young coffee planted on land exhausted by previous planting of coffee. The young trees overbear severely and are particularly short of nutrients because of the poor soil. Both stems and roots die back. *Fusarium oxysporum* infects the roots but it is not clear whether this is a cause of dieback or merely a consequence of the low nutritional status of the plants.

Diseases of Cherries and Beans

Coffee berry disease (anthracnose) (Fig. 8.8)

The fungal agent of this disease had been thought to be a virulent form of *Colletotrichum coffeanum*, but Waller *et al.* (1993) has determined that it is a separate species, *C. kahawae*, related to the group species *Colletotrichum gloeosporioides*, which relationship has been studied by Sreenivasaprasad *et al.* (1993). It attacks young, expanding berries, between 8 and 20 weeks old, and ripe berries. Black sunken lesions are formed, after which the berries rot and fall. Ripe berries can also be affected. Where successive crops overlap, the disease increases in severity. Wet conditions and rainfall favour spore production and dispersal (CBK, 1970b). There are several strains of coffee that are reliably resistant to CBD, but many of the varieties which

Fig. 8.8. Coffee berry disease (*Colletotrichum kahawae*) branches carrying diseased berries (photograph from J.M. Waller, CABI Bioscience).

have been preferred in the past, particularly in Kenya, are very susceptible. Bieysse *et al.* (1995) investigated the genetic diversity and the variability of pathogenic power of *C. kahawae*.

On susceptible varieties, fungicides must be applied during the wet season. Copper fungicides are widely used and chlorothalonil is effective. Resistance has developed to benomyl and carbendazim. Van der Graaf (1992) reviewed the biology, epidemiology and management of this disease. Lambot and Gahiro (1993) found that foliar application of zinc compounds for treatment of zinc deficiency greatly increased the proportion of fruit affected by CBD. The Kenya Coffee Research Station (1993) published revised recommendations for control, including techniques and quantities suitable for smallholders.

Brown blight

This is caused by less virulent strains of *C. coffeanum* in parts of the world that do not have CBD. Anthracnose lesions are formed on berries, leaves and green wood. Ripe berries that are affected are difficult to pulp.

Berry blotch or brown eye spot

This is caused by *Cercospora coffeicola* and can be troublesome in nurseries. It can be controlled by reducing humidity and spraying copper (CBK, 1970b).

Warty disease

This is caused by *Botrytis cinerea*, particularly in wet conditions (CBK, 1970b).

Fusarium stilboides

This fungus can rot the pulp of mature berries.

The last three diseases listed above are not normally very serious but substantial damage can be caused on trees that are overbearing.

Diseases Affecting Nurseries

Damping-off of young seedlings can be caused by any one of *Rhizoctonia solani*, *Fusarium solani* or *F. stilboides*; all can be controlled by applying thiram fungicide to seed or soil.

Hot-and-cold disease

This is a physiological disorder caused by low temperatures that produce a temperature change between day and night exceeding 19°C. It can be particularly troublesome in coffee grown at high altitudes under little or no shade (Mwangi, 1983). Shade of an appropriate density can reduce the temperature variation and hence the incidence of this problem.

Affected leaves are reduced in size and lose colour, becoming yellow to white in an irregular manner, usually starting from the margins; they may be mottled. Affected leaves may eventually scorch and fall. Shoot tips may blacken, distort and die, stimulating excessive branching lower down the stem.

DISEASES AFFECTING ROBUSTA COFFEE

Very few diseases affect robusta coffee. Most selected clones are not very susceptible to *H. vastatrix* (leaf rust). *Hemileia coffeicola*, known as grey rust, attacks many cultivars of *C. canephora* in very wet areas in West Africa. *Fusarium xylarioides* is serious in Angola and locally in places in West Africa. Berry borer (*H. hampei*), which affects coffee in Indonesia, has been controlled by a spray of *Bauveria* fungus. *Cercospora* and *Colletotrichum* (CBD) have no effect at all except for a few clones in central West Africa which get *Colletotrichum* on the leaves.

The roots of robusta coffee are not affected by root diseases, except in Madagascar, where *Clitocybe* spp. have the same effect as *Armillaria*. They must be treated in the same way.

WEED CONTROL

Cramer (1967a) estimated that the losses of potential crop caused by weeds amounted to: Africa 15%, Asia 15%, America 15% and Oceania 16.7%. It is therefore important to keep coffee free of weeds. Dense stands of weeds, such as grasses, can reduce the crop by up to 25%. Coffee is not normally managed in such a way that the ground is shaded to the extent that it significantly reduces the germination of weeds.

Cultural operations help to minimize weeds. Mulch has a significant effect by cooling and darkening the soil. Cover crops discourage weeds; the overall effect can be substantial even if the cover crop is grown only in the centre of the interline.

Specific weed-control operations are always necessary. In the absence of cover crop or mulch, manual or mechanical cultivation between coffee lines will control weeds. There is a risk of damaging coffee roots and frequent cultivation is necessary in the wet season. Damaged roots provide entry points for root diseases.

Herbicides are essential for complete control. They can be used right up to the coffee stems but not too close to foliage which is very close to the ground. The most suitable spray is a mixture of a contact herbicide and a persistent one. Paraquat and diuron mix well together and are frequently recommended for coffee, although diuron can affect coffee on sandy soils. Dalapon is commonly used against perennial grasses. Simazine is often applied to nursery beds before planting seeds. It prevents weeds from germinating. Glyphosate is an alternative contact herbicide with some advantages, particularly a strong effect on grasses and sedges.

Specific recommendations for local conditions and for the use of particular herbicides are available in most coffee-growing regions. Luis and Sanchez (1991) gave practical recommendations for control of weeds in coffee plantations. The CBK (1994) updated recommendations for weed control in coffee in Kenya.

9

BOTANY AND PLANT IMPROVEMENT

BOTANY

The cocoa of commerce comes from the species *Theobroma cacao*. It is one of 22 species in the genus *Theobroma*, in the family *Sterculiaceae*. *Theobroma bicolor* is occasionally used to adulterate cocoa. The species developed in the middle storey of the rain forest of South and Central America. It is diploid ($2n = 20$).

Cocoa will grow to a height of 8–10 m under heavy shade. Grown unshaded, the tree will be less tall. Germination is epigeal, the cotyledon rising to 3 cm above ground level. The first four leaves are held horizontally above the cotyledon. The seedling grows as an unbranched single stem to a height of 1–2 m. Orthotropic growth then stops, and three to five plagiotropic branches are formed. When these have developed, vertical growth will restart, with a new stem growing from the terminal bud. This will again grow for 1–2 m in height. In an unpruned tree, this process may be repeated three or four times before full height is reached. One or more new stems may grow from the base of a mature tree. The vertical stem is known as the chupon and the fan of plagiotropic branches as a jorquette. See Figs 9.1 and 9.2.

Leaves grow only at the growing end of a chupon. Lower down, chupons carry no leaves. These leaves are arranged in a 3/8 spiral; they are symmetrical, with long petioles, which have a swelling (pulvinus) at each end so that the leaves can swivel to catch the light.

Leaves on the fan branches grow alternately in a series of groups, known as flushes. The stem grows rapidly from the terminal bud, producing from three to six leaves, after which stem growth stops. The leaves mature before another flush grows. Fan leaves are very soft and can be any shade between pale green and red. As they mature, they harden and the green colour develops to mask the red. These leaves have shorter petioles than the leaves on chupons and are slightly asymmetrical. They hang downwards, with the upper surface facing outwards from the tree. Flushing requires a plentiful supply of nutrients. These are transferred from older leaves as necessary, to the extent that some may fall.

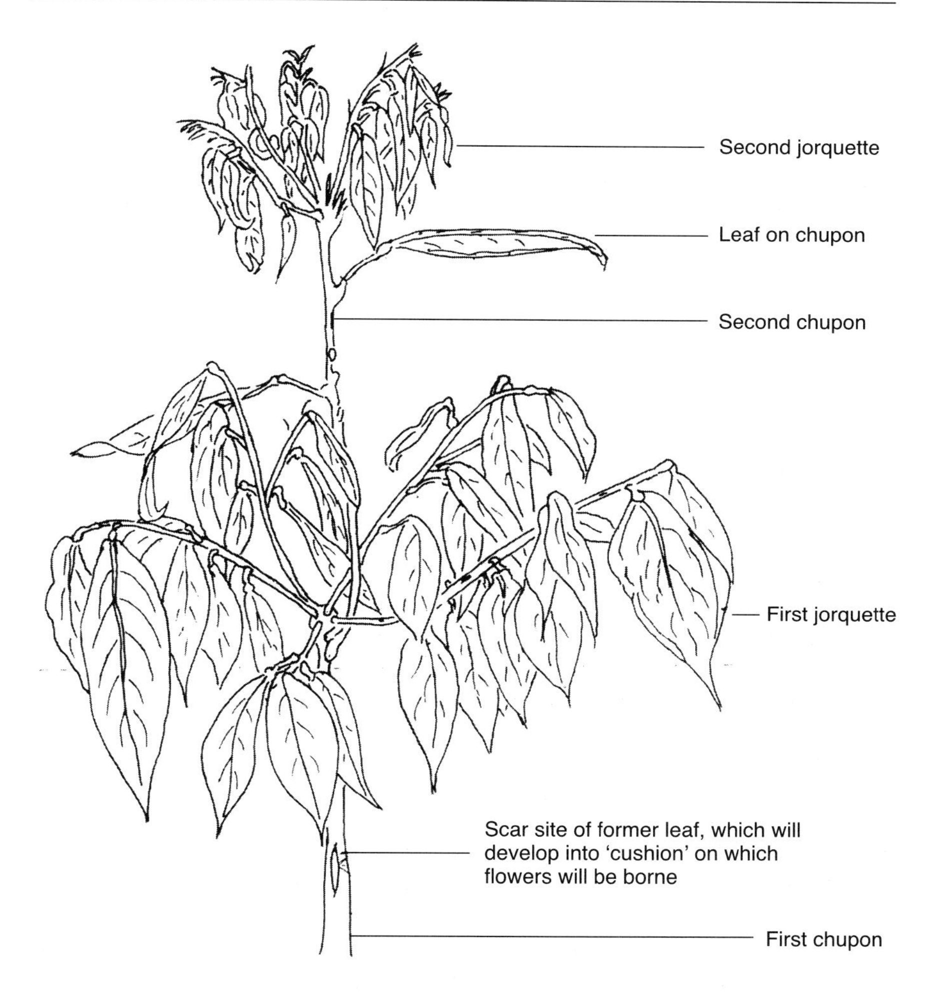

Fig. 9.1. Cocoa tree, showing important vegetative parts.

Hence flushing is sometimes referred to as 'change of leaf'. If nutrients are in short supply, the whole tree may defoliate. The degree of defoliation during a flush is a good indicator of nutrient status.

The seedling forms a tap root which will go to a good depth if the soil is not waterlogged. If the water-table is high, the tap root will only go that far and end in a club. Secondary roots grow sideways close to the surface, so that a mat of roots is formed close to the soil surface. The roots carry a number of mycorrhizal fungi; local fungi in Brazil do not affect mineral nutrient uptake but significant improvements have occurred using introduced fungi

Fig. 9.2. Young cocoa seedling, showing first chupon and first jorquette, Sri Lanka (photograph by K. Willson).

(Kramadibrata and Hedger, 1988). Cuenca *et al.* (1991) reported that *Glomus etunicatum* is the predominant fungus in Venezuela. *Gigaspora* spp. have been effective in Malaysia (Azizah Chulan and Ragu, 1986). Application of fertilizers reduced the level of vesicular-arbuscular mycorrhiza (VAM) infection in cocoa and shade trees.

Cocoa is cauliflorous: the flowers and fruit are borne on the older leafless wood of chupons and jorquettes, each inflorescence arising from a former leaf axil. The site of flowering enlarges with time to form a 'cushion', which will carry many flowers when the tree is mature. The flowers are small (15 mm diameter), hermaphrodite and regular. On pedicels 1–2 cm in length are five free sepals, five free petals, ten stamens in two whorls, one of which is fertile, and a superior ovary of five united carpels. The sepals and the petals, which are smaller, are pink to white. The single style is 2–3 mm in length, ending in five stigmas. Around the style are five longer staminodes in the outer whorl; the inner whorl consists of five fertile stamens (Figs 9.3 and 9.4). Alvim (1984) reviewed the factors affecting flowering.

Natural pollination is mainly by midges; flying females of the genus

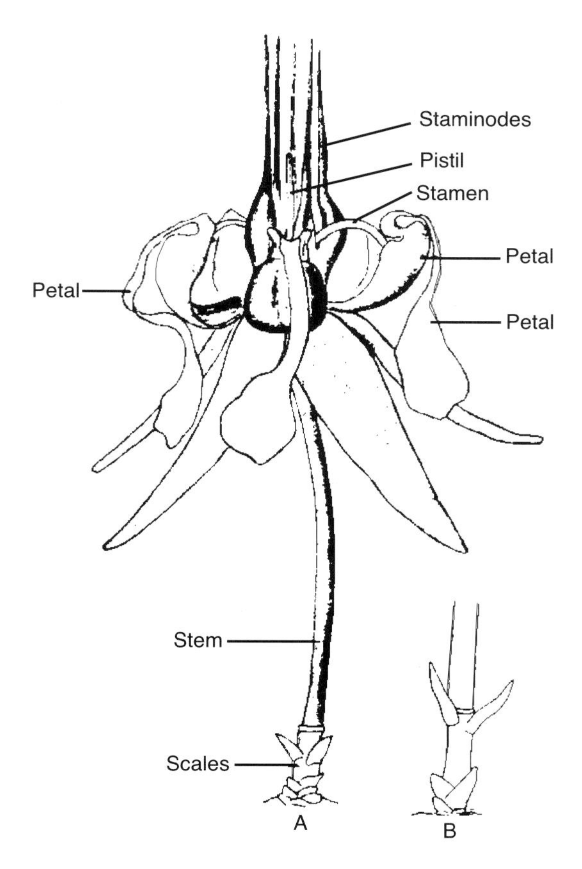

Fig. 9.3. A. Cocoa flower, showing main parts. B. Enlargement of base of flower stem, showing scales. (From van Hall, 1914. Reproduced by courtesy of Macmillan Press Ltd, London.)

Forcipomyia (Winder, 1977). Gabriel *et al.* (1991) reported on the pollinating insects present in cocoa plantations over 10 years old in São Paulo State. Genera and subgenera of Ceratopogonidae (Diptera) were found: 13 *Forcipomyia* (*Euprozoanisia*) spp., three *Dasyheolea* spp., one *Stylobezzia* sp.

Crawling insects, such as aphids and thrips, may also transfer pollen. There is no scent and little colour to attract other insects. The flower structure precludes pollination directly by wind, although the degree of pollination can be increased by creating a wind with a mist-blower (Soria *et al.*, 1986). The wind is presumably moving the insects.

Incompatibility is a major factor limiting pollination; some trees will not set fruit with their own pollen or that from certain other trees. The degree of incompatibility varies from one population to another. The Amelonado variety

Fig. 9.4. Cocoa flowers on stem of tree (photograph from Dr K. Hardwick, Cocoa Research Unit, University of Liverpool).

in West Africa is the only one that is fully self-compatible. Incompatibility arises from non-fusion between gametes in the ovules (Cope, 1962). Other factors affecting pollination have been studied; for example, Aneja *et al.* (1992) studied the effects of carbon dioxide and temperature and Leite *et al.* (1990) compared natural pollination, continuous manual pollination and regular removal of fruits. Aneja *et al.* (1994) reported that carbon dioxide partially overcame self-incompatibility.

Fruit develop after compatible pollination and normally mature in 5–6 months. Small developing fruit are known as cherelles. Up to 80% of the cherelles dry up during the first 3 months. This process is known as 'cherelle wilt', and is a natural thinning arising from competition for nutrients. After 3 months, hormones produced by the fruit prevent wilting (Nichols, 1964).

The fruit is commonly called a pod, although it is considered a drupe (Figs 9.5–9.8). It takes 5–6 months to mature (McKelvie, 1956). It is indehiscent so pods have to be cut off at harvest. In size it is variable – from 10 to 32 cm – in length and its shape can be from cylindrical to spherical. The pod colour can be either of two types: green or green-white immature pods ripen to yellow, while red immature pods darken and may develop some yellow colour. Pods normally contain 20–50 seeds, initially attached to a central placenta. While ripening, the seeds produce an outer layer, which includes a mucilage that is very acid (pH 3.5) and contains citric acid and a high level of sugar. It is therefore an ideal substrate for yeasts. In the final stage of ripening, the outer layer breaks down, releasing the mucilage, which fills the pod and facilitates separation of the seeds. This mucilage is important for the fermentation process, which is an essential part of the development of the flavour of the cocoa. In nature, seeds are distributed by monkeys, rats and squirrels, which gnaw through the outer casing of the fruit and extract the seeds. They suck these to obtain

Fig. 9.5. Cocoa pod (photograph from Dr K. Hardwick, Cocoa Research Unit, University of Liverpool).

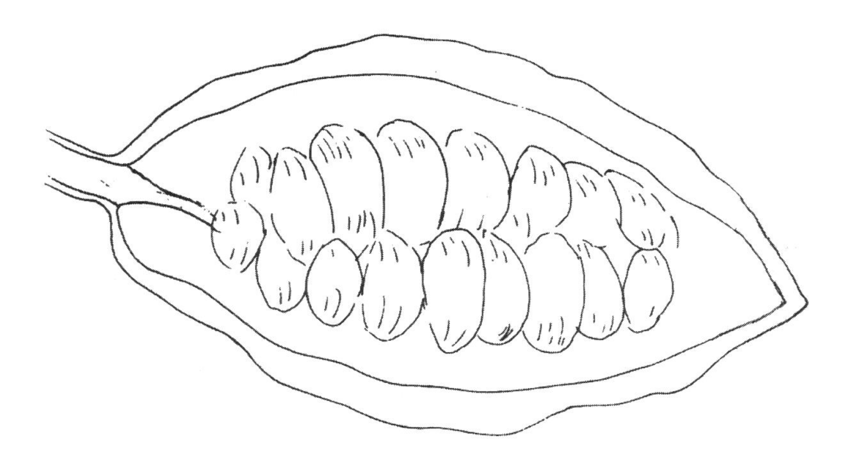

Fig. 9.6. Section through cocoa pod, showing beans.

Fig. 9.7. Mature cocoa carrying pods (photograph from W.R. Carpenter & Co. Ltd, Papua New Guinea).

the sweet mucilage and then discard the seeds; their movements distribute the seeds. The beans contain 55–58% cocoa butter.

There are several distinct varieties of *T. cacao*. Criollo cocoa developed in Central America north of the Panama isthmus. This was the first variety imported into Europe. Pods are long and narrow, externally ridged and either red or green in colour, containing seeds of circular cross-section with white cotyledons. This makes the most flavoursome cocoa. It has been classified as one of two subspecies – *T. cacao* subsp. *cacao*, which is further divided into four forms.

South of Panama the other subspecies developed – *T. cacao* subsp. *sphaerocarpum*. This is known as Amazonian Forastero or Amelonado. It has ovoid pods with ten shallow furrows, green when unripe. The seeds are flattened, with dark purple cotyledons. This variety has less flavour than the Criollo variety.

The two above-mentioned varieties hybridized, initially in Trinidad, to give a series of hybrids known as Trinitario. Individual clones show a wide range of characters, from Criollo type to Forastero type. Usually hardier and more productive than Criollo, the flavour of the best approaches that of Criollo. The genetic diversity of cocoa has been studied by Lanaud *et al.* (1995).

Fig. 9.8. Ripe cocoa pods on tree (photograph from W.R. Carpenter & Co. Ltd, Papua New Guinea).

PLANT IMPROVEMENT

A stand of seedlings includes trees with a wide range of productivity. A relatively small proportion of the trees has been found to produce a major part of the total production (Freeman, 1929). Warren (1992) discussed the possibilities for using modern technology in conjunction with conventional breeding programmes to produce superior planting materials. He argues that the immediate priority is to identify clones with high potential, which can form the starting-point for breeding, using modern technology. The aims of any breeding programme must take into account local plantation management practices, as individual clones will have different responses to management practices, e.g. spacing, as shown by Mooleedhar and Lauckner (1990). The basis of any programme of improvement must therefore be a collection of individuals of known characteristics. To this end, the International Cocoa Genebank was set up in Trinidad (Kennedy and Mooleedhar, 1992). The use of botanical descriptors for characterization of individuals in the collection has been discussed by Bekele (1992). Different clones react variously to different plantation conditions, for example to density of planting (Mooleedhar and Lauckner, 1990) (Fig. 9.9). The breeder has therefore to consider local plantation standards when breeding improved planting material.

Programmes for improvement of planting material have been pursued in several countries. As a first attempt at improvement in Trinidad, seed was taken from high-yielding trees in mixed stands in existing plantations. This approach did not provide the improvement hoped for; the seed was open-pollinated, so only one parent was guaranteed to be selected for high yield (Jolly, 1956). In many investigations, yield was measured by the number of pods per tree, without consideration of the varying sizes of pods and sizes of beans therein. The variation in pod size was quantified as the pod index, the number of pods required to produce 1 lb of dry cocoa. This was found to cover the range from six to 22. Initially, in Trinidad the standard for selection was set at pod index not greater than 7.5 and a yield of 50 pods per annum at a spacing of 3.6 m × 3.6 m. These trees had to be propagated vegetatively, as it was found that seedlings from their seed were less productive than the parent. These clones were planted under differing conditions and it was found that they responded differently to changes in soil and other conditions.

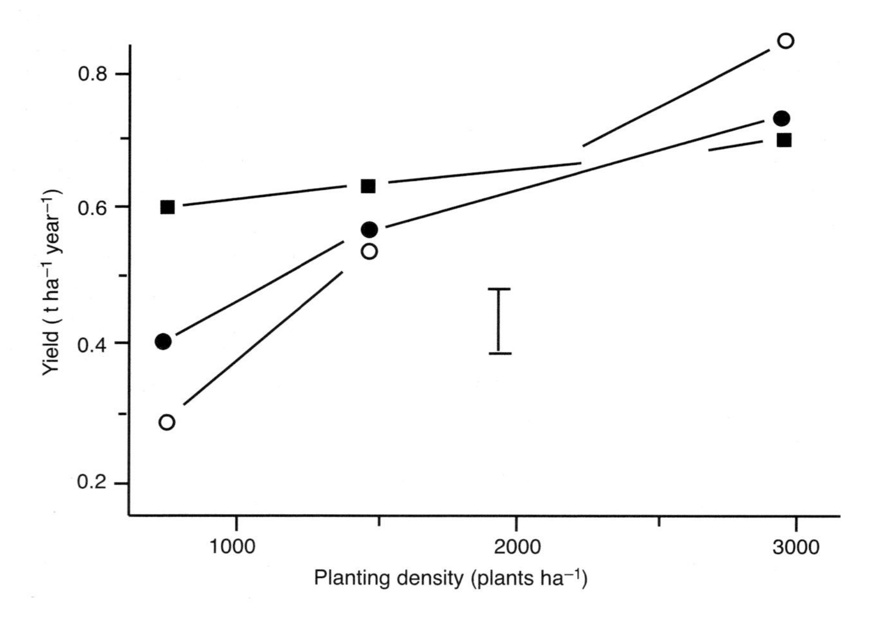

Fig. 9.9. Genotype–environment (GE) interaction among cocoa clones. Differential responses of three Trinidad (TSH) clones to density of stand. Yield of dry beans (Y, t ha^{-1} $year^{-1}$) at three densities of planting (D, plants ha^{-1}). Square symbols, TSH 1 188; dots, TSH 730; circles, TSH 919. The vertical bar is the standard error of the difference between two means. (From Mooleedhar and Lauckner, 1990.)

The spread of witches'-broom disease (*Marasmius perniciosus*) in South and Central America made it desirable to find and incorporate resistance to this disease in the specification for new clones. Two resistant clones were found among trees discovered by expeditions to the forest areas in South America. They produced small beans so they were crossed with some of the earlier selections, which produced some high-yielding clones. The breeding programme continued and it has been possible to incorporate resistance to *Ceratostomella* wilt. This was achieved in three generations by crossing and vigorous selection within the progeny. Selected material has therefore been available in Trinidad since 1945. Initially distributed as rooted cuttings, it was possible to change to hybrid seedlings in 1945 (Bartley, 1969). These selections have been further improved by breeding in resistance to witches'-broom (*M. perniciosus*) and *Ceratostomella* wilt (Freeman, 1969). Sitapai *et al.* (1988) investigated resistance to black pod (*Phytophthora palmivora*). While the objective of cocoa breeding has been, in the first instance, improvement of crop yields, followed by resistance to pests and diseases, the demand for good flavour must not be forgotten. Clapperton (1995) reviewed prospects in the genetics of cocoa flavour.

Ghana is the major producer in Africa. While some trees were imported by missionaries, the most important introduction came from Fernando Po in 1879. Material from the same source was introduced later to Nigeria. As the industry grew, material of Criollo and Trinitario types was brought in and used in crossing and selection. Further introductions were made in 1944 from Trinidad when the spread of swollen-shoot virus posed a major threat. Trees which came originally from the upper Amazon were found to be superior to other material. Crosses were made among these and gave some superior clones, of which ten were released in 1964 and one more later. Apart from the eleventh release, all resulted from hand pollination among the upper Amazon selections. Open-pollinated seeds from plots of these were later released for general planting. Improvement continued by making controlled crosses between the highest-yielding selected material. Apart from yield and quality, resistance to black pod, swollen-shoot virus and drought were also looked for (Toxopeus, 1972). Adu-Ampomah (1996) has reviewed the cocoa breeding programme in Ghana.

Frimpong *et al.* (1995) reported on breeding to improve drought resistance, an important requirement where replanting on degraded soil with diminishing rainfall is necessary.

Improvement by repeated hybridization is a lengthy process. Pinto *et al.* (1990) discussed ways of accelerating the development of new hybrids. Simmonds (1993) displayed in a diagram the optimum programme for cocoa breeding (Fig. 9.10). Warren (1992) discussed the techniques available for plant improvement and their place in improving cocoa in the 21st century.

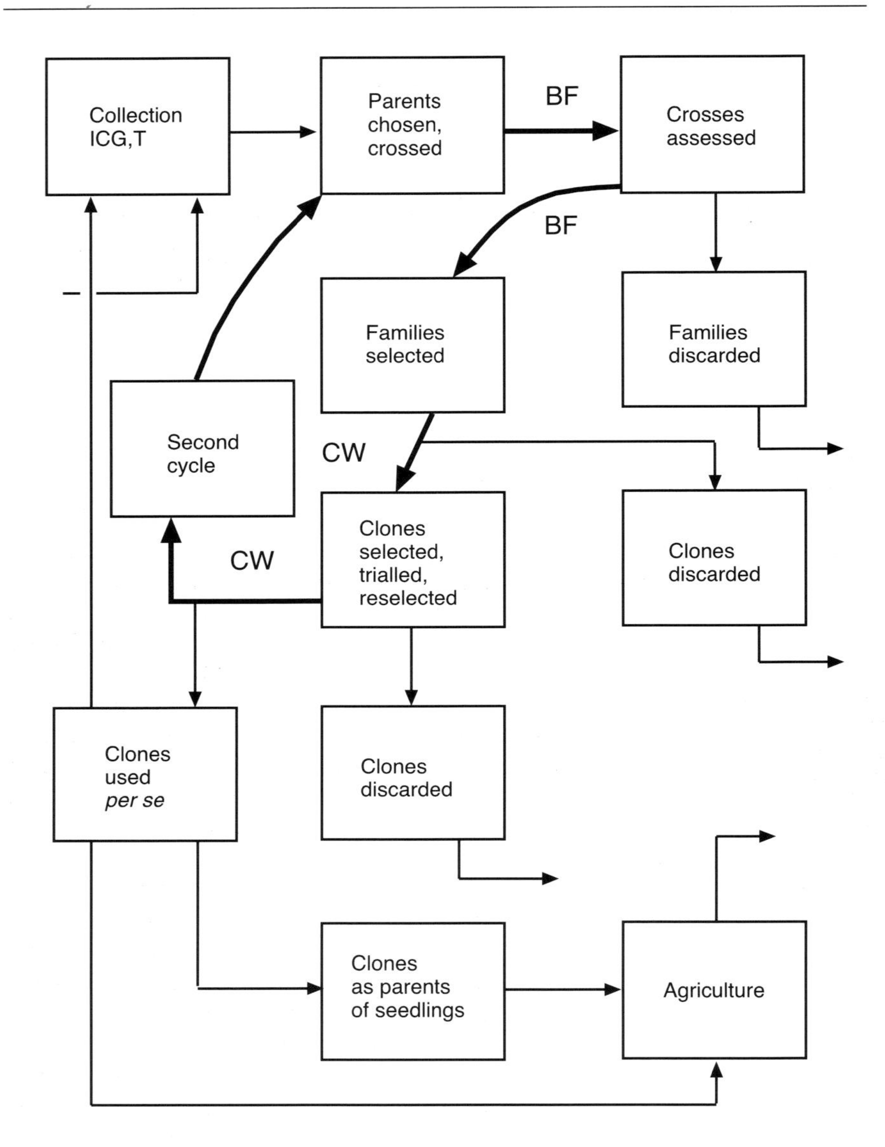

Fig. 9.10. The pattern of cocoa breeding. The cyclical character, emphasized by the heavy arrows, is founded upon alternating selection between families (BF) and between clones within families (CW). ICG, T, International Cocoa Genebank, Trinidad. (From Simmonds, 1993.)

10

Climatic Requirements, Soil Requirements and Management

CLIMATIC REQUIREMENTS

Cocoa developed in the South and Central American rain forest, where the rainfall is high and well distributed, with a short dry season. As a lower-storey tree, the level of insolation is relatively low and is reduced further by cloud cover during rainfall. Temperature is high and varies over a small range during the year. Humidity below the rain-forest canopy is always high.

Such conditions are found only in Equatorial regions. Lower temperatures and the lengthening of the dry season prevent cocoa from being economic too far from the Equator. Over 75% of the world's cocoa is grown within 8° of the Equator. Local variations in climate permit economic cocoa plantations in Bahia, Brazil, 13–18°S, Mexico, Jamaica and the Dominican Republic, 18–20°N, Madagascar, 15°S, and Fiji and New Hebrides, 16–18°S.

The annual rainfall of most cocoa-growing regions lies between 1150 mm and 2500 mm. Rainfall between 2500 mm and 3000 mm is normal in some parts of Central America, Africa, Malaysia and Papua New Guinea. At the lower end of the range, evapotranspiration loss will be close to, and may exceed, the rainfall. Under such conditions, the distribution becomes increasingly important and some irrigation may be desirable unless there is a water-table at a constant high level. Siqueira *et al.* (1995) reported substantial increases in yield when cocoa plants were irrigated. Soils under high rainfall are often poor, due to the greater amount of leaching. Mohd Razi *et al.* (1992) reported the effect of water stress on photosynthesis and growth rates of young cocoa plants.

Most cocoa-growing areas have temperatures within the range 18–32°C. The lowest temperature that will not permanently damage cocoa is 10°C; a monthly average of 15°C is the lowest that can be tolerated. There is no exact figure for a maximum tolerable temperature but it seems that cocoa will withstand temperatures over 31°C for limited periods. Long periods over 30°C affect the physiology of the tree, which loses apical dominance so that axillary buds develop, leaves are smaller but the number of flowers increases.

Figures 10.1 and 10.2 (Alvim, P. de T., 1988) show some relationships between climatic factors and the growth and yield of cocoa trees. In Bahia, Brazil (Fig. 10.1), there is sufficient rainfall each month to ensure that at no time are the trees short of water; leaves grow in flushes at intervals of 3–6 months, irrespective of atmospheric temperatures. Flowering and growth of the trunk, however, only take place when the temperature rises from the minimum levels, around 20°C, which occurs from November to April in the southern hemisphere. In tropical regions, such as Malaysia and Ghana (Fig. 10.2), flowering is concentrated in the months February to July, when the rainfall is heavy; there is no flowering during the dry season. Pods are therefore ripe and ready for harvesting 6 months later, between August and January. Temperatures in these regions never drop below 24°C so this factor is not limiting. The same considerations apply in the tropical Amazon region of South America. Colas *et al.* (1995) reported transpiration rates in cocoa grown alone and under coconuts.

Photosynthesis in cocoa is at a maximum in a light intensity of 25% of full sunlight. Nevertheless, it tolerates extremes of light intensity. It remains alive in heavy forest with a very low light intensity near the ground and, given adequate water, unshaded cocoa will thrive in full sunlight. The stomata remain open and leaves transpire freely (Okali and Owusu, 1975). However, the canopy must be thick enough to maintain high humidity within it. Shortage

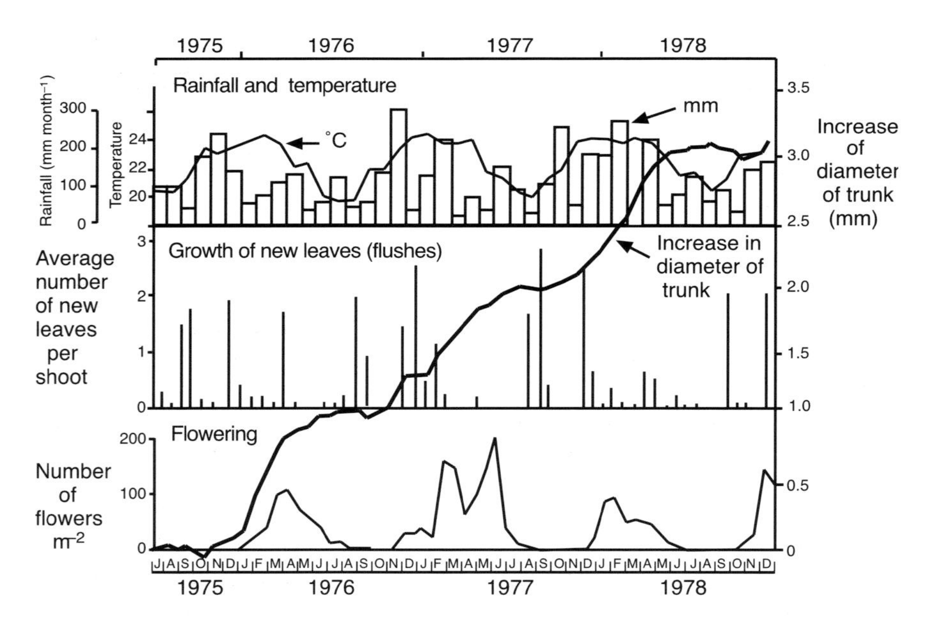

Fig. 10.1. Growth of new leaves and trunk and numbers of new flowers formed in relation to monthly average rainfall and temperature for cocoa in Jucari, Bahaia, Brazil. Data averaged over ten trees. (From Alvim, P. de T., 1988.)

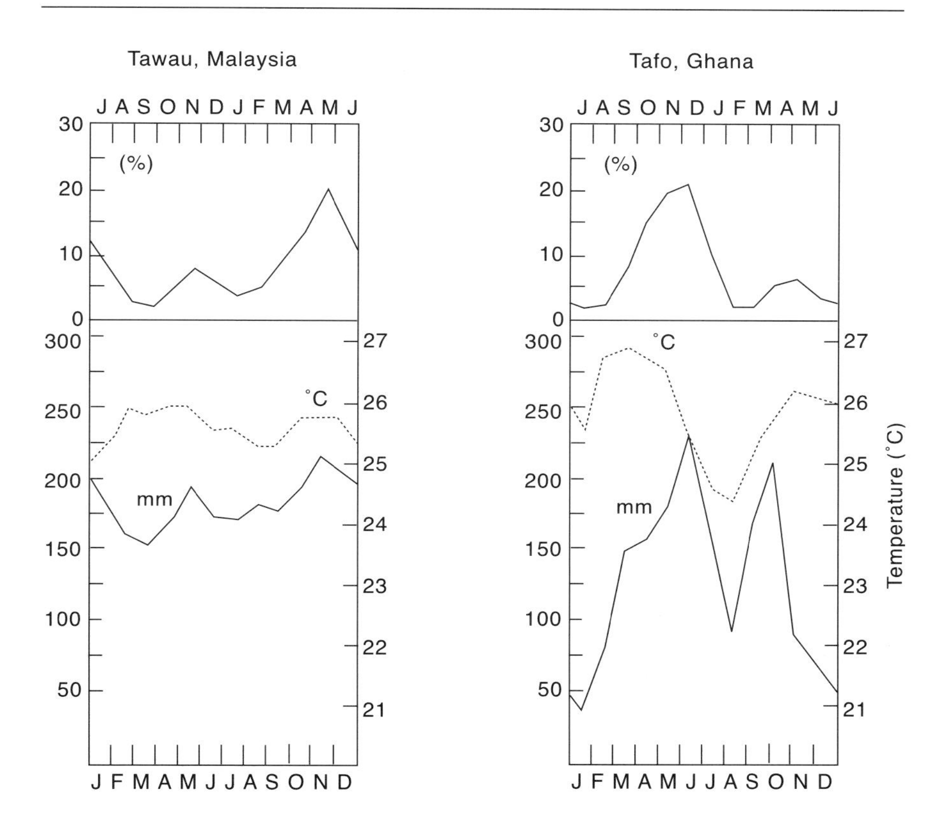

Fig. 10.2. Crop yield, atmospheric temperature and rainfall plotted monthly for cocoa in Malaysia and Ghana (data for 1961). Note: The upper section of the graph is displaced from the lower by 6 months. (From Alvim, P. de T., 1988.)

of water leads to leaf fall and dieback. Nursery plants and young plants in the field need shading to reduce light intensity but all shade can be removed from mature cocoa.

Changes in relative humidity have a greater effect on stomatal resistance than levels of temperature. As relative humidity falls so the stomata close and the resistance increases to restrict transpiration (Balasimha and Rajagopal, 1988). Nevertheless, a period of drought with low relative humidity can cause leaves to wilt. Leaves recover quickly if water becomes available before they fall.

Wind damages cocoa. Leaves are torn, develop marginal and interveinal browning and fall. Cooling by wind induces wider opening of stomata. Shoots are dehydrated by wind. The dry harmattan wind in West Africa sets the limit to the area in which cocoa can be grown. Hurricanes can cause major damage, but trees often recover by producing a new basal chupon.

Dry periods are important in restricting the spread of fungal diseases, particularly black pod (*Phytophthora palmivora*).

SOIL REQUIREMENTS

Soil structure must permit the roots to grow. Soils with too high a proportion of fine particles (fine sand or clay) are therefore not suitable for cocoa, as the spaces between particles are too small. Coarser soils are much better but sandy soils do not retain much water and often lack nutrients. Soils with a high clay content do not drain quickly and become waterlogged when it rains. A soil with roughly equal proportions of sand and clay will have a good structure, with coarser particles leaving free space for roots and retaining a reasonable quantity of nutrients. Cocoa needs such a soil to a depth of at least 1.5 m to permit the development of a good root system. Below that level, it is desirable to have no rocks, hardpans or other impermeable material so that excess water can drain away through the profile. Cocoa will withstand waterlogging for short periods but excess water must not linger around the roots.

Chemically, soil for cocoa should have a pH in the range 5.0–7.5. This is vital for topsoil; lower layers can have a pH outside the range but excessive acidity (pH 4.0 and below) or alkalinity (pH 8.0 and above) must be avoided. In acid soils, aluminium toxicity can be a problem. Exchangeable bases in the soil should amount to at least 35% of the total cation exchange capacity. If the levels of the bases in the top 15 cm of soil are too low, nutritional problems are likely and an appropriate fertilizer programme will be vital. Analyses of soils from some cocoa-growing areas are listed in Table 10.1.

SOIL MANAGEMENT

Clearance of the area to be planted must take due regard of the soil. Land suitable for cocoa most probably carries a stand of forest trees. The forest may be felled and cleared, or the cocoa can be planted under forest shade or a crop tree after clearance of undergrowth and unnecessary trees.

If the forest is to be cleared, this must start at the end of a dry season with cutting and removal of undergrowth. Trees must then be felled during the dry season. Many forest trees are hosts to the fungus disease *Armillaria mellea*, which will transfer to the roots of cocoa. Forest trees should therefore be killed before felling by ring-barking or frilling, which must be done in good time, preferably 2 years in advance. The cut must be inspected regularly, as bark tends to regrow across the gap formed. Dead trees can be felled and carried off the field, or stacked in heaps for burning, which must be done well before the onset of the rains while the timber is dry. The land can be cultivated after removal or burning of the forest material. The ash from large fires can create nutrient

Table 10.1. Chemical properties of some soils in cocoa-growing areas.

Country	pH	Organic matter (%)	Available P (p.p.m.)	Exchangeable bases (mequiv. 100 g^{-1})			CEC (mequiv. 100 g^{-1})	Reference
				K	Mg	Ca		
Brazil	5.1	3.0	5	0.09	3.0	3.9	21.6	1
Ghana	6.4			0.3	7.9	23.7	43	2
Nigeria	7.3		28	2.1	5.4	12.9	25.7	3
Malaysia	5.2		9	0.28	0.5	0.3	13.9	4
Papua New Guinea	6.7	6.8		4.51	4.83	18.2	37	5
	5.3	5.0		0.36	3.86	7.0	17	5

References: 1, Anon. (1975); 2, Brammer (1962); 3, Smyth and Montgomery (1962); 4, Panton (1957); 5, K. Willson (unpublished records).

imbalances; fires should be as small and widely spread out as possible. Heaps of ash from fires should be spread as widely as possible.

Loss of soil must be minimized. Dragging large timber off a field can carry a lot of soil with it; this is likely to be fertile topsoil with a relatively high organic content. Removal of stumps and roots of large trees can create considerable disturbance of the soil.

Roadways should be laid out after clearance of undergrowth but before felling. If drains will be necessary, they should be dug before felling if that is possible. Road construction and drain digging are, in any case, more easily done in dry weather. On sloping land, drains should have a gentle slope from the level contour so that they intercept water moving downhill and minimize erosion. After removal of debris the land needs to be cultivated, followed as soon as possible by planting of a cover crop.

Wood and Lass (1985a) report the benefits of a cover crop; planting a quick-growing cover crop immediately after land clearance minimizes the loss of soil by erosion. Planting cocoa in bare soil leaves all the moisture and nutrients for the cocoa but heavy rain will cause severe losses of soil and nutrients. Mulching is an effective solution but a source of mulching vegetation is needed; cutting, transporting and spreading can be expensive. Most species used as cover crops are legumes, which benefit the nitrogen level in the soil. They need to establish quickly from seed. *Crotalaria striata* and *Pueraria javanica* have been found effective in West Africa. *C. striata* and *Crotalaria anagyroides* are considered too vigorous in Malaysia; *Sesbania punctata* establishes easily and provides shade and soil protection, as it grows to a height of about 2 m and dies out after about 18 months. *Tephrosia candida* and *Tephrosia vogelii* are slow to establish. A mixture of *T. candida* and *C. anagyroides* has been used in Papua New Guinea. Any cover crop that survives the planting of cocoa will be cleared out during weed-control operations.

The lines and positions of shade and cocoa trees should next be marked by stakes, followed by planting of shade trees, which should preferably be well established before the cocoa is planted. The shade trees may give sufficient shelter to the young cocoa, but it may also be necessary to plant a short-lived shelter crop.

When planting under forest shade or a tree crop, the undergrowth has first to be removed and then trees are felled in the forest to adjust the shade to the desired level. The lines and positions of cocoa plants are then marked by stakes; some adjustment of the positions of crop plants may be necessary.

The cocoa is planted when shade and cover crop are established, later in the wet season or in the following year, early enough to ensure that it is well rooted before the dry season. The cover crop may need cutting along the lines of cocoa. Shade and cover crop will shelter the cocoa and minimize soil erosion. The cover crop will ultimately die out or be removed during weed-control operations. Weeds can minimize soil erosion; as the cover of shade of cocoa thickens, weed growth will diminish and the heavy cover will minimize soil erosion by rainfall.

A different sequence of operations has been followed in Papua New Guinea (Blow, 1968). After removal of undergrowth, the permanent shade and cocoa are established under the forest trees, which are then killed slowly and removed individually when they are dead. At no time is the soil left unprotected against erosion, and the cocoa is in production earlier than by following other methods. Falling forest trees may, however, damage shade and cocoa.

11

Field Management

SITE SELECTION

Any site chosen for planting cocoa must meet the criteria for climate and soil discussed in Chapter 10. Slope and drainage must be considered together. Steep slopes will need some form of erosion control, particularly if the site has been clear-felled before planting. After shade, cocoa and cover crop have matured, it is not likely that there will be much movement of water over the surface unless rainstorms are very intense, when drains and terraces will be needed to remove surplus water. Drains may be needed in flat land also, depending on how quickly excess water drains away. Additional drainage may be necessary on flat land at the lower end of a slope, as water often drains downward below the surface and emerges to flood the lower land.

A site, preferably level, is needed for office, stores and most processing equipment. A sun-drier with movable canopy covers a significant area. Fermenting boxes can conveniently be sited on a slope. Good access is needed to the office and other buildings, with roads for estate transport to all fields. Morris (1997) discussed the design and maintenance of estate road systems.

A supply of clean drinking-water is necessary. A nursery needs to have water available throughout the year.

It is important to evaluate the level of wind on any potential site. Strong, steady winds can damage cocoa severely; heavy damage to leaves leads to defoliation. Violent winds, such as hurricanes, can break trees, but these usually recover by growing a new chupon. Shade trees minimize damage to cocoa by wind; unshaded cocoa should only be planted where it is known that strong winds never occur or the site is well sheltered by hills. Seasonal winds, such as the harmattan in West Africa, prevent cocoa planting in otherwise suitable areas. The extent to which such winds affect an area which is otherwise suitable should be known before a site is finally chosen.

PROPAGATION

Most cocoa is raised from seed as the raising of seedlings is easier and cheaper than any vegetative method. Vegetative propagation is used only when plants of specific clones are required. Seedlings produce a chupon first, so that seedling plants have a normal pattern of growth. Vegetative propagation needs single-leaf cuttings or a bud from a leaf axil; the majority of leaves on a cocoa tree are on the jorquette and the first growth from such cuttings is a jorquette. From a cutting, therefore, the first jorquette starts at ground level. From a budded plant, the jorquette starts at the budding height, which is not very far from the ground. In either case, the low jorquette is a nuisance; it interferes with harvesting, weed control and other operations. Pruning is required at an early age to produce trees with a normal configuration, which are easier to work.

Seed Gardens

Trees of proven quality must be planted in seed gardens to ensure that the material produced is good. The clones that are to be planted should have been selected on the basis of the results of trials on the progeny. Selected clones must be self-incompatible. The selected trees are propagated vegetatively; often only two parents are used in a seed garden.

The planting arrangement in the garden depends on whether pollination is natural or by hand. If pollination is allowed to occur naturally, the ratio of one parent to the other should be 1, provided that both parents are self-incompatible. If one is partly self-compatible, it is used only as a male parent. Fewer trees of this clone will be planted and its pods will not be taken as seed. A distance of at least 200 m between a seed garden and other cocoa will prevent unwanted pollination.

Hand-pollination enables plants of chosen crosses to be produced, and no others. Pollination is simple. The work must start early in the morning; the white anthers darken during the day and are useless if discoloured. The first part of the operation is to collect male flowers with white anthers and put them in a small container. They must be used on the day of collection. Selected female flowers must also be freshly opened. At the tree to be pollinated, an anther must be removed from a male flower using forceps. Check that the anthers are open; they should be dull white, while unopened ones are shiny. The anther is gently brushed against the style to transfer the pollen. It is sometimes necessary to remove one or two staminodes to gain access to the style. Pollinated flowers are marked by a piece of thread. It is not normally necessary to cover flower buds to prevent natural pollination – fewer than 5% of pods are from natural pollination – but this must be done if crosses of guaranteed origin are required for breeding investigations. Up to 800 pollinations can be made by one person in 1 day if flowers are plentiful; this will probably give up to 300 pods.

Pollination by hand can be done so that seed is available at the most convenient time for nursery planting. All cherelles and young pods on the trees at the time of pollination must be removed, as they will compete with the pollinated flowers. The mechanism of cherelle wilt is prevented if a number of pods are set at the same time. If more pods are set than the tree can support, the pods will be smaller and the tree may defoliate and die back. Hand-pollination permits more than one pair of clones to be planted in a seed garden. A series of selected crosses can then be made and unwanted ones avoided.

Germination

Seed will germinate and produce good plants when taken from pods not more than 15 days underripe; younger seed will not produce good plants. At least 90% of seed will have germinated within 15 days. Removal of mucilage by rubbing beans in sawdust or sand and washing will reduce the germination time to 7 days. This operation may be necessary to minimize insect attack. Germination in wet sand or under wet sacking can avoid wasting nursery space on blind seeds and may produce more uniform plants. It is probably more costly and care must be taken to plant seeds before the germ has reached a length of 1 cm. Normal practice is to plant the seeds directly into the soil in plastic bags.

Nurseries for Seedlings

A nursery site needs to be level, convenient for a supply of suitable soil and for the fields to be planted. A reliable supply of good water is vital; a usage of the order of 10 litres per seedling watering should be allowed for. Watering will be necessary at least twice a day. Nurseries must be shaded. As nurseries are normally only used for a few years, a temporary structure is adequate. Locally cut poles support wires, on which palm fronds or other vegetation form the shade. The shade must be heavy when seeds are sown and will be slowly thinned as the plants develop, so that the shade density is the same as that in the field when plants are ready. Successful nurseries have been made under rubber, oil-palms and *Gliricidia*.

Plants are now almost invariably raised in black polythene bags; the size of bag depends on the length of time the plants will remain in their nursery – from 25 cm × 18 cm for 3 months to 40 cm × 25 cm for 6 months or more.

A good loam topsoil is ideal for filling the bags. Sand can be added to a heavy soil to lighten it. Cocoa shell can be added without detrimental effects but most other possible diluents are not beneficial. Erwiyono and Goenadi

(1990) found that a mixture of 25% fine coconut husk with 75% sand was effective. The filled bags are arranged in beds, supported at the sides by strips of bamboo, if necessary. Beds 75–100 cm wide permit easy inspection and access. Initially, bags are touching but growth is better if they are separated a little as they grow.

Seed should be planted not more than 1 cm deep. If planted too deep, the cotyledons, which are pushed above the soil surface in the epigeal growth, will not emerge properly. The plants do not need a lot of attention in the nursery, apart from watering, which must not be overdone, as this may promote attack by *Phytophthora palmivora* or anthracnose (*Colletotrichum gloeosporioides*). Copper fungicides control these diseases. Insects can be controlled by insecticides, either on the ground around the beds or on the plants. Weeds are not usually a problem; hand removal is all that is necessary. Diuron has been used successfully without damaging the cocoa when weed growth was excessive.

Using a fertile topsoil, there is no benefit in applying fertilizer. A poor, acid topsoil benefits from limestone, 15 g per bag of soil. Poor growth on infertile soil can be improved by applying nitrogen, as sulphate of ammonia or urea. Foliar fertilizers can be used, but with care; cocoa is very sensitive to fertilizer toxicity. Plants can be kept in the nursery for up to 6 months. The length of time is dictated by the availability of seed and the rainy season, which is best for planting out.

Cocoa seedlings are difficult to move over long distances. Alvim *et al.* (1982) found that bare-root seedlings, raised in soil beds, not in bags, are best. All but two leaves are removed; the remaining two are treated with an antitranspirant.

Vegetative Propagation

Cuttings

There is considerable variation in the ability of cuttings to form roots. Within each type of cocoa, there is variation from one clone to another. Amazon clones generally root easily; Amelonado are more difficult. Trinitario is very variable, while Nacional de Ecuador is very difficult. In addition, variations occur according to the time of year; in some types, the degree of success falls away markedly in the dry season.

Special gardens of mother trees have been planted in Trinidad, for example, at 2.0 m × 2.0 m. These are shaded with bananas and *Gliricidia* to give 20–50% light. Heavy fertilizer applications are necessary. The number of cuttings produced by each tree varies considerably from one type to another. Cuttings are taken almost entirely from jorquette branches; they should be taken at the semihardwood stage; the upper surface of the shoot should be brown.

Glicenstein *et al.* (1990) found that, if 5-year-old clonal trees (chupons with apical sections removed) were bent over and the apex tied down, the buds on the upper surface produced multiple chupons. These gave many more orthotropic shoots from which cuttings could be made than the unbent trees.

Nurseries for cuttings

Cuttings are usually rooted in sand, coconut-fibre dust, composted sawdust or any similar material with a low content of organic material. After rooting, they are transplanted into a potting soil, which is usually based on the local field soil.

For many years, cuttings were raised in closed bins – boxes containing either rooting medium or potting soil over a layer of stones or gravel for drainage. The boxes were sealed with glass at the top. The bins stood under a dense overhead shade, and the light intensity was further reduced by covering the glass with newspaper or cloth. The cloth had to be chosen carefully to transmit the optimum level of light. The paper or cloth was kept wet to reduce the temperature. Cuttings were rooted in bins containing rooting material; 21 days is normally allowed for this. The rooted cuttings were transferred to bins containing potting soil. The amount of shade was gradually reduced until the young plants were in ambient conditions; this usually took at least 2 weeks (Moll, 1956a).

Bins have now been replaced by individual pots (Moll, 1956b). Almost invariably nowadays, the pots are made of black plastic (polythene) 35 μm thick, with a diameter of 12 cm and a height of 25 cm. They are filled with potting soil around a central core of rooting medium. It is sometimes possible to fill the pots entirely with soil, to which sand has been added. The pots are placed in double rows under shade, leaving space between the rows for access for watering and maintenance. The pots must be watered regularly to keep them moist (Fig. 11.1).

Alternatively, pots may be covered by polythene sheet, which is sealed to the ground at its edges. Shading reduces the light to 15% of the normal daylight. The pots are well watered before sealing; thereafter, all that is necessary is a light watering every third day.

Cuttings can have from two to five leaves and must include one or two buds. They are taken from the mother trees early in the morning and the leaves cut to about half their length. They are then dipped in a hormone rooting solution and planted as soon as possible. Evans (1951) reported that the optimum hormone treatment is to dip the cut end into a solution of 8–10 g of equal parts of α-napthaleneacetic acid and β-indolebutyric acid in 1 l of 50% alcohol. Roots will fill the pots after about 4 weeks. Thereafter, the polythene sheet can be removed slowly, increasing the aperture every few days. When the polythene is fully removed, the overhead shade is reduced slowly until the level of light is similar to that in the field. Plants in pots can be taken to the field without disturbing the roots.

Budding

This method has been used at times on a limited scale. The seedling rootstock may modify the performance of the budded clone, improving a poor clone but

Fig. 11.1. Nursery of clonal cocoa plants under wire netting, which would have carried shade of palm leaves at an earlier stage (photograph from W.R. Carpenter Co. Ltd, Papua New Guinea).

reducing the performance of a good clone. The compatibility of scion and rootstock must be checked before budding on a large scale. Bud wood should be taken from a tree just after the leaves have fallen. Rootstocks should be about 4 months old.

The usual method of budding is either a form of patch budding or an inverted-T budding. In the former, a small square of bark is lifted after cutting an inverted U. A matching piece of bark including the bud is placed under the flap. In inverted-T budding, a shield of bark, 4 cm long, carrying the bud, is inserted in the vertical cut of an inverted T below the flaps of bark on the stock. In either case, the bud is bound tightly in place with raffia, waxed tape or clear plastic strip to prevent loss of moisture. The binding can usually be removed after 2–3 weeks. Buds may remain dormant but can be stimulated by removing a piece of bark slightly wider than the wound about 8 mm above it. When the bud is growing, the stock can be cut off above it.

Grafting

This has little practical application. Both saddle and wedge grafting methods are effective, and side-grafting on to a susceptible rootstock has been used to detect virus infection in bud wood. Sreenivasan (1995) described a method of grafting on to 3-week-old seedlings.

Yow and Lim (1994) reviewed green-patch budding and side-grafting in detail.

Marcotting

This has been used to propagate from older material. A strip of bark 7.5 cm wide is removed and the area covered with sawdust in a polythene sheet. Roots will form and the branch can then be cut off for planting out.

Propagation *in vitro*

Early workers with this technique found that cocoa explants produced excessive amounts of callus and were loath to produce shoots. Adu-Ampomah *et al.* (1990a) were able to produce somatic embryoids from cotyledons and developed a method for their development into plantlets. The same workers (Adu-Ampomah *et al.*, 1990b) devised a method for conversion of shoot-tip explants into plantlets. The effects of γ-irradiation were also reported. Improved methods have been reported by Söndahl *et al.* (1993).

Callus will form on staminodes and leaves under suitable conditions, but oxidation of the polyphenols inhibits the differentiation and growth of embryos in the callus. However, by using the technique of temporary immersion (Teisson, 1994), some callus cells will grow into plantlets. Berthouly *et al.* (1995) compared somatic embryogenesis from flower buds with the same from leaves. Figueira and Janick (1993) reviewed the use of micropropagation for germplasm conservation.

SHADE

The natural habitat of cocoa is under shade and much commercial cocoa is grown under shade (Figs 11.2 and 11.3). Mature cocoa, however, will give a high yield if grown unshaded, provided that adequate moisture and nutrients are available throughout the year. Young cocoa needs shade of an appropriate density to produce trees of the desired shape and geometry. Too little light induces long internodes and few branches, while too much light induces too much branching and short internodes, which produces bushy trees. As the level of shading is reduced, the level of response to applied fertilizers increases. High levels of fertilizer under heavy shade are counterproductive, reducing yields (Murray, 1954; Cunningham and Burridge, 1960). There is a choice between shade, low fertilizer and low yields and no shade, high fertilizer and high yields. The effect in the long term is shown by Ahenkorah *et al.* (1987) (Fig. 11.4). The level of shade required is therefore related to the fertilizer levels proposed. Heavy shade is necessary for young trees but this will have to be thinned or removed during the first few years to provide the optimum shade level for mature cocoa.

A shade tree must therefore give a heavy, even shade throughout the dry

Fig. 11.2. Young cocoa plants in sleeves under leguminous shade (photograph from W.R. Carpenter Co. Ltd, Papua New Guinea).

Fig. 11.3. Young hybrid cocoa under coconuts (photograph from W.R. Carpenter Co. Ltd, Papua New Guinea).

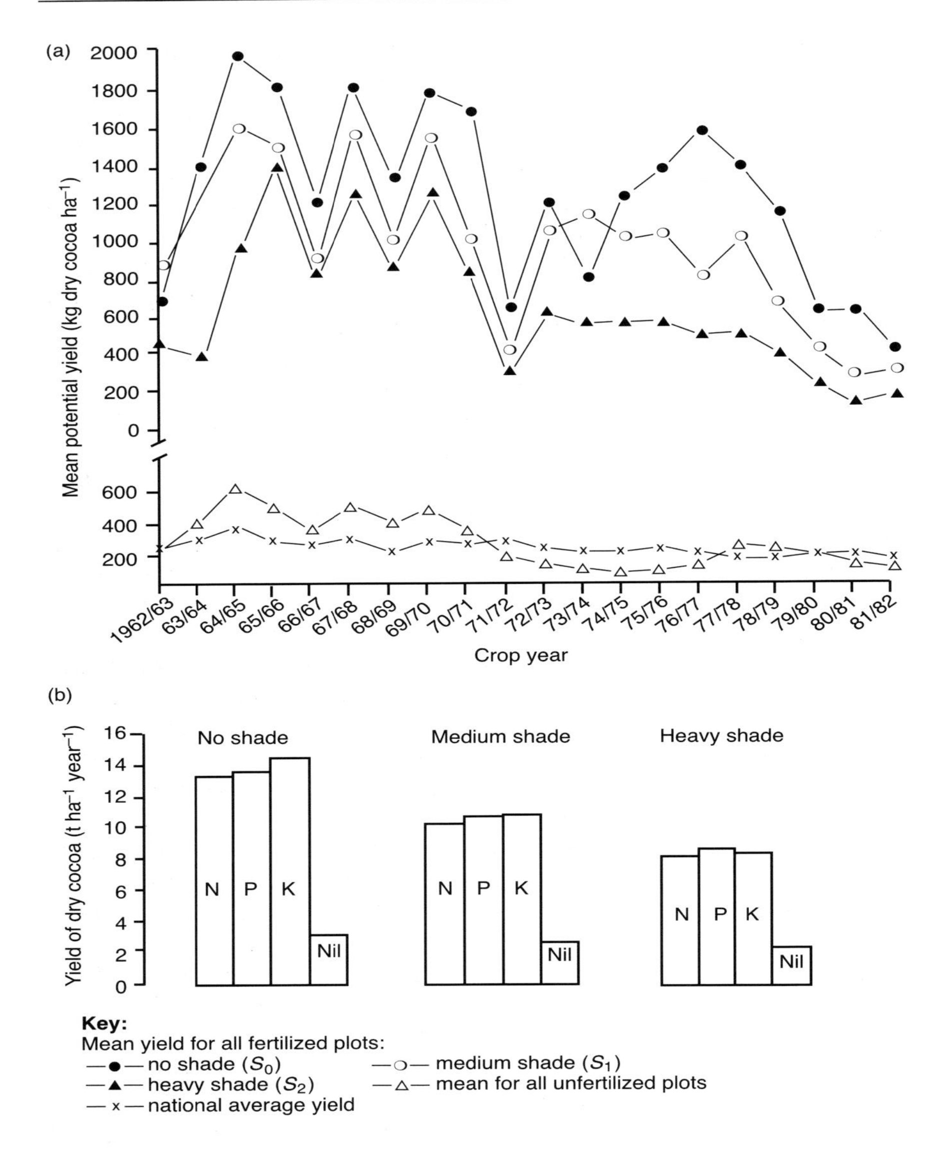

Fig. 11.4. Effects of NPK fertilization and levels of shade on the yield of cocoa (from Ahenkorah, 1987).

season. It must also be possible to thin the trees or remove them at a later stage, without causing undue damage to the cocoa. A number of species are in common use as shade trees. *Gliricidia sepium* ('madre de cacao' in Central America) is a legume that is frequently planted as temporary shade. It establishes easily from planting long stems. Its light foliage falls in the dry season, but if the trees are lopped earlier the new stems which follow will retain their leaves. *Leucaena glauca* and *Leucaena leucocephala*, also legumes, are widely used. They are easily established in well-drained soil by sowing seed in the field, but cannot compete with weeds. The common varieties seed profusely, so volunteer plants can become a weed problem. They have been planted in hedges along cocoa interrows or in clumps. Invasions of defoliating beetles, which moved on to the cocoa, have reduced the popularity of these trees. Strains that produce less seed and some sterile hybrid clones are now available, but are not widely used. Laup (1994) reported that, in Papua New Guinea, problems arose from pests transferring from *Leucaena* to the cocoa, so that *Gliricidia* is now preferred.

Some species of *Albizzia* spread and carry light feathery foliage, which gives a good shade, but are not widely used in cocoa. These are also legumes and propagation is from seed; germination is poor unless the seed is scarified with hot water or acid. *Albizzia moluccana* and *Albizzia falcata* are damaged by strong winds; *Albizzia chinensis* is less brittle but is more difficult to establish and grows more slowly. Several *Erythrina* species, also legumes, are widely used. *Erythrina poeppigiana* (mountain immortelle) and *Erythrina glauca* (swamp immortelle) are widely used in the Caribbean, but severe attacks of diseases have developed in recent years. *Erythrina lithosperma* (thornless dadap) is widely used in Sri Lanka and Indonesia as conventional shade. In Samoa it is used as a ground cover in unshaded cocoa, being kept cut to a height of only 75 cm. Dadap loses its leaves in the dry season but new growth, which grows following lopping, will retain its leaves. *Parkia javanica* is a tall leguminous tree that is easy to establish. It provides little shade in its early years. Several other species are used on a limited scale. The number of tree species with the appropriate characteristics is limited. In some cases, one type of shade, e.g. *Gliricidia*, is planted for the young cocoa, to be replaced by another, e.g. *Albizzia* or a commercial crop, as the cocoa approaches maturity.

Shade trees need to be established before the cocoa is planted. Both temporary and permanent shade must therefore be planted as soon as possible after land clearance. Positioning of temporary shade depends on the species. Often a clump around each cocoa site is appropriate. Permanent shade should be within the cocoa lines. The distance between shade trees depends on the species and the level of shade required; 12–18 m apart would include most commercial situations.

PLANTING IN THINNED FOREST

This method is commonly used in West Africa. The natural forest is thinned to leave about five very large trees and 35–45 smaller trees per hectare. The cocoa is planted after removal of unwanted trees and undergrowth. The lines of cocoa are likely to be irregular. By this method, cocoa can be in the ground 1 year sooner than by clear felling. The irregular planting creates difficulties for mechanical maintenance, but the method is appropriate for small farmers. It cannot be used where the dry season rainfall will not support both established shade and cocoa.

COCOA WITH OTHER COMMERCIAL CROPS

Cocoa has been tried with other commercial crops which could provide ground cover, temporary or permanent shade. The profitability of the land could be increased.

The most successful partner for cocoa is coconut, which provides an appropriate level of shade and does not compete severely for nutrients (Fig. 11.3). The soil must be suitable for cocoa; coconuts grow well on many soils totally unsuitable for cocoa. The cocoa is planted under mature coconuts. The dense mat of coconut roots must be broken up before planting cocoa either by ploughing or by digging very large holes for the cocoa plants. Daswir *et al.* (1988) found that cocoa under coconuts was not as profitable as cocoa under *L. leucocephala* shade; however, Smith (1985) reported that pest and disease problems were less severe under coconuts.

Cocoa has been tried under oil-palm, rubber, areca (betel) nut, cola, peach palm and nutmeg. These crops at their normal spacings provide too heavy a shade. Planting under old rubber which has been thinned to a very wide spacing has been successful. Bananas are used successfully as temporary shade for young cocoa. The bananas must be removed before the cocoa grows too large.

Some commercial mixed-cropping agroforestry systems have been described by Dubois (1987). Alvim (1990b) reviewed possible mixed-cropping systems. Tan *et al.* (1991) reported that there was an economic benefit from replacement of 86% of *Gliricidia maculata* shade in cocoa by *Carica papaya* (pawpaw). In addition to the pawpaw fruit the cocoa crop increased, as a higher proportion of the cocoa trees bore fruit. Vaast (1990) described an effective system for combining cocoa with annual food crops. Joseph *et al.* (1993) reported on the management of a combination of coconut, black pepper, cocoa and pineapple.

WIND-BREAKS

In situations where strong winds occur frequently, cocoa must be protected by the planting of dense rows of trees as wind-breaks. A wide variety of species has been used. Mango (*Mangifera indica*) or galba (*Calophyllium antillanum*) are planted in Grenada and Jamaica. *Dracaena* spp. or *Hibiscus* spp. are used in Trinidad. In Fiji, cloves (*Eugenia aromatica*), malacca apple (*Eugenia malaccensis*), mahogany (*Swietenia macrophylla*) or mango are used. Teak (*Tectona grandis*) is used in Samoa and cinnamon (*Cinnamomum zeylanicum*) in Zanzibar. Herzog (1994) reported that smallholders in Côte d'Ivoire plant cocoa under a mixture of trees, most of which are used traditionally for a variety of purposes.

SPACING

Closer spacings usually produce higher yields in the first years after planting. The advantage decreases as the trees age. Close spacings can restrict access to mature trees, but a closed canopy is achieved more quickly, which helps to minimize weed growth and soil erosion. Interlocking canopies on closely spaced trees maintain a high humidity, which favours the spread of diseases. Close spacings are more expensive to plant. Mooleedhar and Lauckner (1990) investigated yields from three clones at three spacings. All clones gave higher yields per unit area at the closer spacings. There was an interaction between clones and spacings: one clone was significantly more productive at the closest spacing. The closest spacing, 1.8 m × 1.8 m, was below the customary minima in the main growing regions, shown below. On the other hand, Redshaw and Zulnerlin (1996) found no significant difference of yield in 4-year-old seedlings between 816 and 1633 seedlings ha^{-1}. In practice, spacings between 2.5 m × 2.5 m (1600 plants ha^{-1}) and 5 m × 5 m (400 plants ha^{-1}) are used. In each growing region, there is a narrow range of spacings which are customarily used – for example, 4.6 m × 4.6 m to 5 m × 5 m in Papua New Guinea, Sri Lanka and Samoa, 4 m × 4 m in Central America, 3 m × 3 m to 4 m × 4 m in South America and 2.5 m × 2.5 m in West Africa. Trees in thinned forest in West Africa cannot be planted in exact geometrical positions because of the random distribution of forest trees, but the resulting density is usually close to 1600 ha^{-1}, equivalent to a rectangular planting at 2.5 m × 2.5 m.

COVER CROPS

A full discussion of cover crops is included in Chapter 10. Where a cover crop has been planted after land clearance, holing will clear the crop from each plant site. Control may be necessary to prevent the cover crop from smothering young cocoa plants, but it may provide some shade. Development of the shade and cocoa trees will ultimately shade out any cover crop. Subsequent weeding operations will remove any surviving plants. A cover crop will not normally survive longer than 18 months.

MULCH

Mulching is not a common operation in cocoa; cutting, carrying and spreading mulch is an expensive operation. Mulch minimizes erosion and breaks down to yield some nutrients without competing with the cocoa for nutrients and water. Mulch not infected with *Phytophthora* pod rot may help in controlling this disease by providing a barrier between the trees and the reservoir of infection in the soil.

FIELD PLANTING

Seed-at-stake

It is common in West Africa for seed to be planted directly in the field. Often, more than one seed is planted at each position so that there is ultimately more than one tree per position. Only a very small hole is necessary for planting a seed. Multiple sowing ensures that there is at least one tree at each position. Losses due to poor germination and damage to young plants are of little consequence.

Planting Nursery Plants

A hole must be dug at each position of a depth to avoid bending the tap root. The diameter must permit insertion of the pot or allow for spreading the secondary roots of bare-rooted plants. Often a larger hole is dug. Fertilizer, usually animal manure, is often mixed with the soil from the hole when it is available. The size of the holes varies and is usually within the range 40–60 cm diameter or square, with a depth between 25 cm and 75 cm. In many soils, it is advantageous to dig holes several months in advance to allow the soil to weather.

Planting should be done in favourable weather, when the soil is wet; preferably at the beginning of the wet season. Plants should be well watered before removal from the nursery, kept under shade and planted as soon as possible. The soil around the roots should not be disturbed in transit. Shade conditions in the field should be similar to those in the nursery. If field shade is significantly lighter, young plants should be shaded temporarily, for example by palm fronds.

If nursery plants have to be transported a long distance, they should be packed with bare roots and reduced to two leaves, which are themselves cut to about half-size. They should then be sealed in plastic to prevent moisture loss.

PRUNING

This operation has two major objectives.

1. To get trees into the most convenient shape for maximum production and ease of harvesting and maintenance.
2. To achieve the best control of pests and diseases.

The above together should maximize the return from the trees. Seedlings should develop a jorquette at an appropriate height. This height varies significantly from tree to tree. It has also been found that the height of jorquettes decreases as the strength of the light increases. The density of shade can therefore affect the height of jorquettes. If a jorquette is considered too low, it can be cut off. New chupons will grow from buds on the original chupon; the strongest can be selected and all others removed. In due course, the second chupon will produce a jorquette at a higher level. Vegetatively propagated plants generally form a jorquette at ground level. It may be possible to remove this after a chupon has grown and a second jorquette has formed at a more convenient height.

After formation of a jorquette, access can be improved by removal of branches which droop very low. Removal of jorquette branches, reducing their number to not less than three, opens up the trees, allowing more light to enter and decreasing the humidity within the canopy. There is no firm evidence to prove that this minimizes black pod infection. Removal of too many stems reduces yield. Benefits from pruning were reported by Sena Gomes *et al.* (1995).

Trees should be inspected regularly. Dead, diseased and badly damaged wood should be removed.

A second chupon may be permitted to grow and form a second jorquette at a higher level. There is no clear evidence to show whether such two-storey trees are more or less productive than single-storey trees.

Prunings are often left in the field to rot down. This returns the nutrients they contain to the soil for reuse. Lim and Vizhi (1993) showed that they could be converted to charcoal. This would, however, remove their nutrients from the field; they would have to be replaced by purchased fertilizer. Diseased prunings should always be removed. In some situations, control of diseases requires the removal of all prunings.

INFILLING OR SUPPLYING

There will usually be some losses when a field of cocoa is planted. Losses should not, in normal circumstances, exceed 15% of the trees and are more usually around 10%. It is customary to plant new trees to replace the losses. Thereafter, as the majority of trees mature, it becomes more difficult to supply vacancies and tend isolated young trees. After the first year, supplying should be restricted to large areas, which may arise, for example, as damage following the fall of a shade tree.

POLLINATION FOR CROP PRODUCTION

The output of mature pods is dependent on the degree of pollination of the flowers. Cherelle wilt will prevent the production of mature pods beyond the capacity of the tree when the crop is heavy, but when pollination is inadequate wilt will abort more pods than is necessary to reduce the crop only to the capacity of the tree, so the crop will be lower.

Cocoa grown in countries where the pollinating midges (*Forcipomyia* spp.) were not indigenous yielded poorly, due to lack of pollination. The degree of pollination depends on the number of midges. Elizondo (1988) found that the number of midges increased when banana pseudostems were placed by the cocoa trees. The midges bred in the banana stems; there were significant differences between banana varieties in the numbers of breeding midges.

Muller *et al.* (1988) found that manual pollination was much more effective than the natural insect pollination in the first year. Heavy pollination (40 flowers per tree on alternate days) and light pollination (ten flowers per tree on alternate days) increased yields by 500% and 200% respectively. However, in the next year, the heavily pollinated plants carried so many pods that they produced few flowers and suffered more severely from physiological wilt. The lightly pollinated plants continued to give a yield approximately double that of insect-pollinated trees.

Forced ventilation, using mist-blowers, has been employed to improve natural pollination. Soria *et al.* (1986) reported a substantial increase in pollination following the use of a mist-blower.

HARVESTING

The length of time from fertilization of a flower and harvesting a ripe pod is affected by the temperature. As the temperature rises, the rate of maturation increases. Dependent on temperature, the ripening period can vary between 4.5 and 7 months. The amount of crop available for harvest at any one time varies.

Temperature variations throughout the year will tend to concentrate crop in the periods of higher temperatures. Nevertheless, there is usually some crop available at other times. The amount of flowering and the number of pods set are higher at periods of high ambient temperature. There will therefore be an increased number of pods to harvest several months after the period of high temperature (see Fig. 10.1, p. 112).

Pods must be harvested when fully ripe (see Fig. 9.8, p. 107). Beans from unripe pods produce low-quality cocoa. Ripe pods should be removed as soon as possible; once ripe, they are more likely to be attacked by fungal diseases or by animal pests (monkeys, etc.), which extract the sweet ripe beans from the pods. Harvesting should be carried out regularly at intervals of 10–14 days, which may be extended to 3 weeks. Ripe beans can germinate inside the pod. Germinated beans do not make good cocoa. Pods can be kept for a few days between harvesting and opening. Such a delay has been found to be advantageous; the fermentation heats up more quickly and the proportion of undesirable 'purple beans' falls with delays of up to 4 days.

Pods must be cut off the tree without damaging the cushion, on which further fruits will form. Special knives are used – a machete or cutlass for pods within an arm's reach and a knife on a pole for higher pods. At each harvesting round, sufficient pods must be cut to completely fill one or more fermenting boxes. Partly filled boxes do not ferment properly (see Chapter 19).

Pods then have to be opened for removal of the beans. Sometimes, they are collected together in the field and opened there; only the beans need transporting to the factory. Otherwise, the pods are transported to the factory and opened close to the fermenting boxes. Pods are usually opened with a knife but this risks damaging beans. As an alternative, pods can be cracked on a stone or triangular piece of wood or by hitting with a piece of wood. The beans are removed by hand from the pod shell and put to ferment. At the present time, almost all the pods produced in the world are opened manually. The placenta to which the beans are attached should not be removed with the beans.

With a view to mechanization in the future, the breaking behaviour of cocoa pods has been investigated (Maduaho and Faborode, 1994). A mathematical model of cocoa pod breakage has been devised (Faborode and Dinrifo, 1994). A dehulling and winnowing machine has been described (Ademosun, 1993).

The husk is a waste product. If it is left or spread on the soil in the plantation, it rots down and returns nutrients to the soil. This is not permissible where diseases or pests can infect the cocoa treas. Oduwole and Arueya (1990) investigated the possibility of extracting potassium for fertilizer from the husks and showed that it would be viable. In West Africa, the ash from burning husks, containing about 40% potassium hydroxide, is a major source of potash for soap manufacture. Adomako *et al.* (1995) described improvement of the ashing process on a large scale. Cocoa husk can also be used as feed for cattle (Rodrigues Filho *et al.*, 1993) or sheep (Adomaho and Tuah, 1988). Biogas production from husks was investigated by Bopaiah *et al.* (1988) and ethanol production by Samah *et al.* (1992).

12

MINERAL NUTRITION AND FERTILIZERS

NUTRIENTS REMOVED IN CROP

The exact quantity of nutrients removed in the cocoa crop depends on the nutritional state of the tree. On average 1000 kg of cocoa beans will contain about 30 kg of nitrogen (as N), 8 kg of phosphate (as P_2O_5), 40 kg of potash (as K_2O), 13 kg of calcium (as calcium oxide (CaO)) and 10 kg of magnesium (as magnesium oxide (MgO)). Nutrients are also required to build the framework of the tree. Further quantities of nutrients will be removed in the shells of the pods, which are particularly high in potash. In many cases, these are allowed to rot down on the plantation so that the nutrients return to the soil. The distribution of these nutrients is likely to be uneven, as pods are concentrated in localized areas for opening.

The quantity of fertilizer required will depend on the amounts of nutrients available in the soil. Fertilizers may not be necessary at all in the early years on cocoa planted into a rich soil, whereas cocoa on a relatively infertile soil will need fertilization from planting onwards. Rosand (1995) investigated the nutritional requirements of cocoa under different production systems.

EFFECT OF SHADE

The level of light reaching the canopy of a cocoa tree has a large effect on the crop yield and the demand for fertilizers. With a low light level, under heavy shade, the crop yield will be low. Application of fertilizer will have little or no effect on crop yield. With a high light level, under very thin or no shade, yields will be higher than under heavy shade. In this case, there will be a substantial response, in terms of increased crop, to applications of fertilizer. High light levels with low nitrogen availability can produce symptoms of nitrogen deficiency. The need to maintain the balance between the various nutrients means that phosphorus (P) and potassium (K), at least, must increase as nitrogen is increased. The benefits of fertilizer application and reduction of shade were demonstrated by Ahenkorah *et al.* (1987) (see Fig. 11.4, p. 126). Adeyemi (1995) reported benefits from NPK fertilization of rehabilitated cocoa plantations.

EFFECTS OF INDIVIDUAL NUTRIENTS

Nitrogen

This is a major component of all parts of the tree and is particularly necessary for the production of the vegetative components. The regular shedding and regrowth of the leaves of cocoa requires a continuous availability of nitrogen in the soil, as does the growth of fruit, which occurs throughout the year (Jadin, 1980).

Phosphorus

This nutrient is vital for the growth processes. The quantity required is relatively small. Its availability in soil can be restricted by chemical fixation, particularly in acid soils. Regular application of a small quantity of phosphatic fertilizer is very important. In most soils, incorporation of phosphate in planting holes gives a significant improvement in early growth.

Azizah Chulan and Ragu (1986) found that vesicular-arbuscular mycorrhiza (*Gigaspora* spp.) significantly increased the uptake of phosphorus. Very high levels of phosphorus application depressed mycorrhizal development.

Potassium

Vital for the mechanism of growth, potassium is particularly important for fruit production. Fruit contain a large proportion of this nutrient. Jadin (1980) discussed the importance of potassium. Regular application of substantial quantities of potash fertilizers are necessary on many soils to maintain crop levels. On soils with high levels of potassium, this element will interfere with the uptake of calcium and magnesium, causing deficiencies. The fertilizer programme must then provide calcium and magnesium but exclude potassium (Sugiyono, 1992).

Calcium

This is another nutrient that is essential in a small quantity. On soils of the optimum pH range, a deficiency is very unlikely. Very acid soils would have to be treated with lime, which would add calcium to the soil.

Magnesium

This nutrient is essential and is a constituent of chlorophyll. Deficiencies of magnesium can occur on some soils, so that magnesium fertilizers are often required. The balance between calcium and magnesium is important; excess of one can limit uptake of the other.

Zinc

This element is only required in very small quantities but is, nevertheless, essential. Deficiencies occur quite frequently and must be corrected. Nakayama (1989) established optimum application rates of zinc and boron fertilizers and the corresponding levels of these elements in the third leaf.

Boron

Deficiencies of the element are not uncommon, although only small quantities are required. A deficiency can have a severe effect on the yield and health of a tree, so correction is important, but in excess the element is toxic.

Aluminium

This element can be available in large quantities in acid soils. De Santana and Cabala-Rosand (1984) found a negative correlation between cocoa growth and aluminium level in the soil. Sugiyono (1989) found that saturation levels of aluminium above 30% reduced the uptake of phosphorus and calcium and inhibited root growth. In such cases, enough lime should be applied to neutralize all the aluminium.

Chlorine

Excess of this element has an adverse effect on a cocoa tree. Care must be taken when planting cocoa close to the seashore; the salinity of the soil must be checked and a wind-break provided to prevent salt-water spray from reaching the cocoa. Potassium should always be applied to cocoa as potassium sulphate, not muriate of potash, in order to avoid chlorine toxicity.

MINERAL FERTILIZERS IN COMMON USE

Nitrogen

Sulphate of ammonia is freely available. It has a relatively low nitrogen content (*c*. 20%). It acidifies the soil, so that continuous use will reduce the soil pH and soil acidity may become a problem; increased application levels of potassium and magnesium may become necessary. Urea has a much higher nitrogen content (*c*. 45%) and is therefore much cheaper to transport. It should not be applied in hot, dry weather, as it will lose nitrogen.

Phosphorus

Phosphorus is usually supplied as a superphosphate. Single superphosphate contains about 20% P_2O_5, and also sulphur. Double and triple superphosphates contain no sulphur and 46–53% P_2O_5.

Potassium

Potassium sulphate is the most appropriate for cocoa. It contains about 50% K_2O and 32% sulphur. Muriate of potash (potassium chloride) is not suitable for cocoa because of its high chlorine content (about 47%).

Calcium

It is not normally necessary to apply calcium to cure a deficiency, but lime or limestone may be necessary on very acid soil to raise the soil pH.

Magnesium

Kieserite is an impure form of magnesium sulphate and is quite acceptable as a fertilizer. Pure magnesium sulphate (Epsom salt) is much more expensive. It may be used as a constituent of compound fertilizers.

Zinc

This nutrient is usually applied as a foliar spray, as very large ground applications are necessary to cure a deficiency. Zinc sulphate is available as monohydrate or heptahydrate. Either can be used. It is soluble in water and ideal for foliar sprays. Zinc oxide and oxychloride are not soluble in water, but either is available in a form that makes a suspension in water, which is suitable for a foliar spray.

Boron

Borax is soluble in water and is normally used in a foliar spray.

ORGANIC MANURES

Where organic manure is available, its use on cocoa is likely to be beneficial. The most common is that from animal enclosures. The organic material has a beneficial effect on the soil, improving the structure and its capacity to hold water and nutrients. The content of mineral nutrients is usually variable and low, so that addition of some mineral fertilizer is desirable. Chepote (1995) reported a reduction in mineral fertilizer use and increased cocoa yields following application of organic materials.

In most soils, manure mixed with the soil in planting holes benefits young plants, increasing early growth.

On older trees, mulching with animal manure or cut vegetation is beneficial. It improves the soil, provides some nutrients and discourages weeds. It is, however, an expensive operation so is often not viable economically.

SYMPTOMS OF NUTRIENT DEFICIENCY AND TOXICITY

- *Nitrogen deficiency*. The leaves are reduced in size and lose their colour. The areas between the veins become lighter green and then yellow to almost white as the deficiency worsens, The veins remain darker but do lighten as the deficiency progresses. There is a colour photograph in Wood and Lass (1985b).
- *Phosphorus deficiency*. The plants become stunted, with the leaves, especially young ones, getting smaller. Mature leaves develop paler margins and tips, while young leaves become paler between the veins. Scorching of leaf margins develops later. Young growth has short internodes and leaves at an acute angle. The stipules often remain after leaves have fallen.
- *Potassium deficiency*. Interveinal areas close to the leaf margins turn pale yellow and then necrotic. The necrotic areas, with a yellow leading edge, advance towards the centre of the leaf. There is a coloured photograph in Wood and Lass (1985b).
- *Calcium deficiency*. This is very similar to potassium deficiency. The necrotic areas near the leaf margin join up quickly to form a continuous

marginal necrotic area. The unaffected area along the veins develops a wavy edge. There is a coloured photograph in Wood and Lass (1985b).

- *Magnesium deficiency.* This is also similar to potassium deficiency. The interveinal areas become very light in colour, with the veins outstandingly dark. Necrosis at the leaf margins spreads quickly and advances between the veins behind a yellow strip. Necrotic patches form ahead of the marginal area.
- *Sulphur deficiency.* All of the leaves on the plant become yellow, in blotches on older leaves. Young leaves are initially bright yellow, later becoming pale yellow-green like older leaves. Sumbak (1983) reviewed the information on sulphur deficiency and fertilization.
- *Iron toxicity.* A pale yellow area develops on both sides of the midrib on older leaves. Necrotic areas develop around wounds and spread rapidly. There is no necrosis of the tip and margin.
- *Iron deficiency.* The leaf laminae lighten in colour and a network of veins appears against a background that develops to be almost white. There is a coloured photograph in Wood and Lass (1985b).
- *Manganese toxicity.* Young mature leaves develop irregular pale green to yellow areas, sometimes with some necrosis of the veins.
- *Manganese deficiency.* The younger leaves turn yellow and develop a chlorosis around the main veins. The tip and margins later scorch.
- *Zinc toxicity.* Young leaves become olive-green or develop pale green patches.
- *Zinc deficiency.* The young leaves develop red veins and distort severely. Leaves become very narrow, out of normal proportion. The leaves often develop a wavy edge and distort to a sickle shape. Chlorotic patches develop in rows parallel to the veins.
- *Copper toxicity.* Young leaves become dark olive-green with prominent veins and they pucker down the centre. Younger leaves develop pale green areas, randomly distributed.
- *Copper deficiency.* Young leaves become smaller, without change in colour. Leaf tissues at tips collapse and later form a brown edge.
- *Boron toxicity.* The older leaves develop marginal scorch and necrotic areas, while younger leaves twist downwards, with interveinal chlorotic areas and green veins; tips and margins become necrotic.
- *Boron deficiency.* Young leaves are small, pale in colour, thick and brittle; they harden in a backward curve or a spiral twist. Older leaves are unaffected. Boron-deficient soils severely restrict the growth of young plants of certain cultivars (Chude, 1988).
- *Aluminium toxicity.* Older leaves develop paler green to yellow areas in the interveinal areas close to the leaf tip. These areas scorch but the symptom spreads very slowly. Sometimes the interveinal areas of older leaves, near the base of the leaf, blacken.

- *Chlorine toxicity*. Interveinal areas near leaf margins become pale yellow. These quickly fuse and scorch. The scorched areas spread, more quickly in interveinal regions. Unscorched areas become dark green or grey.
- *Molybdenum deficiency*. Young leaves become thin and translucent, and later produce mild chlorotic mottling. Some marginal scorch develops later.

ASSESSMENT OF FERTILIZER NEEDS BY CHEMICAL ANALYSIS

Foliar analysis is used successfully in many crops as a basis for calculating fertilizer requirements. Unfortunately, it is not as successful in cocoa because the leaves have a relatively short life. When a new flush of leaves grows, older leaves usually senesce and die. The amounts of the various nutrients in the leaves vary throughout the short lifespan of each flush. Saleh (1973) investigated the relationships between leaf age and nutrient content. It is therefore almost impossible to take samples of leaves which are on each occasion of comparable physiological age. Snoeck (1984b) reviewed the use of plant analyses as a guide to fertilizer requirements.

It has, however, been found that soil analysis can provide a reliable guide to fertilizer needs. The Institut de la Recherche sur le Café, Cacao, Thé et Autres Plantes Stimulantes (IRCC), in Côte d'Ivoire has developed a standard method for calculating fertilizer needs from soil-analysis data (Jadin, 1976, 1980). Jadin and Vaast (1990) reported on the fertilizers required to rehabilitate plantations in Togo. They concluded that the calculated fertilizer rates would produce economic benefits.

FERTILIZER APPLICATION LEVELS

The level of shade has a large influence on fertilizer needs and applications. Under very heavy shade, there is very little response to fertilizer, so that application is not viable. Unshaded cocoa gives a large response to fertilizer, provided that the trees are well managed and there is adequate moisture in the soil throughout the year. At least 1300 mm of water (rainfall plus irrigation) is recommended for the full response by unshaded cocoa.

On poor soils, particularly on soil from which old cocoa has been removed, fertilization must start with planting-hole application of 200 g of superphosphate mixed in the soil. On such soils, application of nitrogen, phosphate and potash fertilizers should start very soon after planting. On richer soils, several years may elapse before fertilizer application becomes necessary. From the start, the application must be of nitrogen, phosphate and potash. For example, a suitable mixture could be made from two to four parts of sulphate of ammonia, two parts single superphosphate and one part sulphate of potash. The amount of sulphate of ammonia should vary according to the level of shade. Sulphate of potash should be used rather than muriate because cocoa is sensitive to chlorine. Alternatively, a compound fertilizer, such as 6:5:6 or 10:5:10 NPK, can be used.

Application rates should start at around 25 kg N ha^{-1} in the first year after planting and increase steadily over the first 5–6 years to about 150 kg N ha^{-1}. The exact level will depend on the degree of shading and the quality of the soil. In most areas, local advice will be available.

Most nutrients can be applied in foliar fertilizers, a method that is particularly effective for the minor nutrients. Malavolta (1986) discussed foliar fertilization in Brazil.

13

PESTS, DISEASES AND WEED CONTROL

Cocoa is at risk from many pests and diseases which thrive in the essential warm, humid climate in which it is grown. Cramer (1967b) estimated the annual loss of cocoa crop to be 588,000 t from pests 368,000 t from diseases and 337,000 t from weeds. These add up to 44.9% of potential production in that year. This proportional loss of crop from these three factors is higher than for any other major crop in the world (Wood, 1990).

Major pests and diseases have not followed cocoa around the world, as has happened with many other crops. In each new growing area, major pathogens have transferred from indigenous hosts. Diseases which appear the same in different continents have been found to be caused by different species; for example, while *Phytophthora* pod rot is caused by *Phytophthora palmivora* in the centre of origin, the Amazon region of South America, it is caused by *Phytophthora megakarya* in West Africa and in Bahia, in the south of South America, *Phytophthora capsici* is the major cause. The worldwide spectrum of cocoa pests and diseases has been reviewed by Keane (1992).

MANAGEMENT OF PESTS AND DISEASES

With the wide range of pathogens ready to invade cocoa simultaneously at most locations, control to permit an economic yield of cocoa must embrace all aspects of management. In the first instance, resistance to the pathogen will eliminate the need for further control measures. Simmonds (1994) reviewed recent data supporting the view that there is horizontal (polygenic) resistance to cocoa diseases. He concluded that there is now evidence of horizontal resistance to five pathogens. The only commercially successful use of resistant material has been in Papua New Guinea, where vascular-streak dieback (VSD), *Oncobasidium theobromae*, has become a minor problem. The heterogeneous mixture of cocoa planted in Papua New Guinea gave a wide range within which to look for resistance. It may well be more difficult to find, in more homogeneous cocoa plantings, resistance that can be used for breeding.

Biological control is environmentally attractive and should be cheap in the long run if a stable population of predators can be established. The use of 'crazy ants' (*Anoplolepis longipes*) to control mirids, *Amblypelta* spp. and *Pantorhytes* spp. is the only widespread example at the present time, but the habit of ant colonies to move on discourages some growers.

Horticultural management within and around the plantation is extremely important. Regular sanitation is vital; all diseased material must be cut off and removed, some pests can be removed by hand and pods can be enclosed in tubes to foil podborer, while borer channels and cankers can be painted with the appropriate chemicals. Tree spacing and control of tree size by pruning can enhance the effectiveness of sanitation and provide an environment within the plantation that minimizes the spread of pathogens. Removal of indigenous vegetation close to the plantation can remove alternative hosts. Maintenance of an adequate level of shade helps to control mirids, although absence of shade assists aeration of the cocoa canopy, which discourages some diseases. The choice of shade tree can be important; tall shade discourages mirids, *Pantorhytes* spp. and *Glenea* spp.; coconuts are ideal and are used extensively in some countries. However, some potential shade or crop trees can be alternative hosts to cocoa pathogens. Smith (1985) found that pest and disease problems in Papua New Guinea are more serious under *Leucaena leucocephala* shade than under coconut palms. Ground cover – plants or mulch – helps to minimize *Phytophthora* pod rot.

Pesticides are still vital if cocoa is to crop at a profitable level. Some insecticides that are unpopular environmentally must be used as the alternatives have disadvantages. For example, γ-HCH is very effective on mirids; its volatility compensates for any inaccuracy in spraying. Some effective alternatives are unpleasant to handle; propoxur is effective but costly. The use of pesticides has been studied in several countries and optimum treatment levels of permitted chemicals established (for example, Khoo *et al.*, 1983).

With each growing region having different major pathogens, quarantine control must be rigorously enforced. Diseases can be contained effectively; within Papua New Guinea, several islands where VSD did not exist have been maintained free of this disease.

PESTS

Cramer (1967b) estimated the annual loss of the potential cocoa crop from pests to be 5.5% in America, 15% in Africa, 8.3% in Asia and 10.5% in Oceania. The number of insect species that have been reported as feeding on cocoa exceeds 1500 (Entwistle, 1972). Introduction of cocoa to an area where it had not previously been planted often leads to some local pests attacking the introduced crop, although they had not previously been reported on cocoa. Some of these adapt to cause significant damage. Most of the pests reported do not create any significant problems but there are sufficient major pests to cause significant crop loss.

Idowu (1995) investigated the benefits of intercropping with food crops and showed that certain food crops could be valuable in integrated pest management. Williams (1995) reported that basic data for integrated pest management for cocoa had been entered into a database available on the World Wide Web.

Mirids or Capsids (Fig. 13.1)

These sap-sucking bugs cause severe damage in many countries. Their eggs are buried in the epidermal layer of chupons, jorquette branches, pods and pod stalks and hatch in up to 17 days. Five successive nymphal stages follow before the winged adult appears.

Where the insects have fed by sucking sap lesions are formed which turn black. Lesions in unhardened stems cause wilting and ultimately death. They also allow injurious fungi to enter. The extent of losses of pods due to myrid feeding varies from one country to another; losses from stem damage are serious everywhere. Cros *et al.* (1995) reported that the attractiveness of specific cocoa trees to mirids is related to the amount of flavonol present in the tree.

These insects are usually 7–11 mm long and are of various, mainly dark,

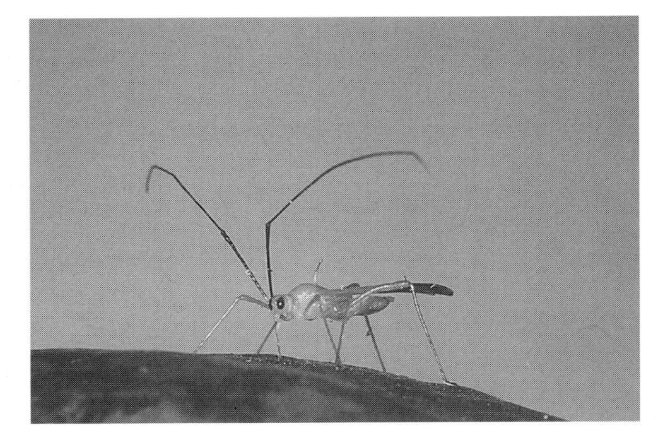

Fig. 13.1. Cocoa mirid (*Helopeltis* sp.), Uganda (photograph from International Institute of Biological Control).

colours. The important species in particular countries are: *Sahlbergella singularis* from Zaïre to Sierra Leone, *Distantiella theobroma* in Côte d'Ivoire and Nigeria, *Monalonion* spp. in South and Central America, *Helopeltis* spp. in Sri Lanka, Indonesia, Papua New Guinea and Malaysia, *Pseudodoniella* spp. in Papua New Guinea, *Boxiopsis madagascariensis* in Madagascar and *Platyngomiriodes apiformis* in Sabah.

Control of mirids has traditionally been by application of cyclodiene organochlorine insecticides, particularly benzene hexachloride (BHC), but resistance to BHC and hence other compounds of this type has developed within the last 20 years. Some carbamate insecticides are seen to be effective alternatives. Application, by mist-blower, should be most effective when related to the life cycle of the insects. Thus, in West Africa, applications should start when it is known that the population increases if there is no control. This occurs in July–August; further applications are made at approximately monthly intervals until the end of the year, when an uncontrolled population reaches its maximum. In some countries, a degree of secondary control can be achieved by ants: *Anaplolepis longipes* (crazy ants) in Papua New Guinea, *Dolichodercus thoracicus* (cocoa black ant) in Indonesia (Ooi, 1992) and *Wasmania auropunctata* in Cameroon (Braneau de Mire, 1969). *Sahlbergella* is also parasitized by *Euphorus sahlbergellae*. A small wasp, it can infect up to 30% of the *Sahlbergella* in the dry season.

Shield Bugs

Antiteuchus (= *Mecistorhinus*) *picea* and *Antiteuchus triptera* occur in South and Central America and Trinidad. They are minor pests in nurseries in Trinidad; their effect on mature cocoa has not been evaluated.

Bathycoelia thalassina is a large green bug that attacks cocoa between Zaïre and Côte d'Ivoire. While their eggs are laid mainly on leaves, less often on stems, they feed mainly on pods, which inhibits development of beans and makes pods abort.

Resistance to organochlorine pesticides has reduced their effectiveness in controlling shield bugs. Control can be achieved by following an application of lindane by an application of thiodan, unden or basudine 28 days later.

Leaf Hoppers

Empoasca devastans in Sri Lanka, *Affroccidens* spp. in Ghana and *Chinaia rubescens* in Costa Rica cause leaf-hopper burn: distortion and premature fall of leaves. In South and Central America, *Horiola picta* attacks cushions, pods and stems and may cause pods to wilt. These species are more active in unshaded cocoa. Fenitrothion or formothion will control them.

Psyllids

Tyora (or *Mesahomatoma*) *tessmanni* is important in Africa from the Congo to Sierra Leone. These small insects lay eggs in vegetative buds, young leaves, flowers and young fruit and form fluffy white colonies on young stems and flower cushions. Buds may die and flowers and fruit atrophy (Kaufmann, 1975). These insects may take part in pollinating cocoa.

Aphids

In Africa, *Toxoptera aurantii* is the most important species, causing leaves to crinkle and fall and flowers to wilt. Attacks are rarely serious; menazon gives effective control.

Scale Insects and Mealybugs (Fig. 13.2)

Scale insects within the genus *Asterolecanum* are known throughout the tropics. They are minor pests on cocoa, their main importance is that they produce stem swellings, which can be confused with swellings caused by cocoa

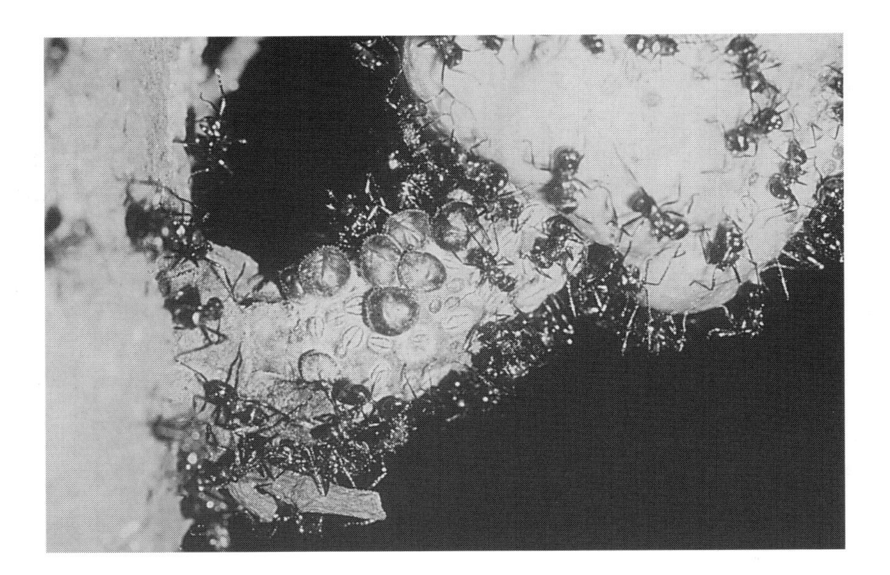

Fig. 13.2. Ants on an infestation of mealybugs on cocoa (*Planococcus citri*); note the white wax covering of the mealybugs, Papua New Guinea. (Photograph from International Institute of Biological Control.)

swollen-shoot virus. *Asterolecanum* spp. swellings can be recognized by dimpling, known as the 'pit-and-gall' syndrome. *Stictococcus* spp. infest pods heavily and damage them. They are confined to Africa.

Over 40 species of mealybug, within the superfamily *Coccoidea*, infest cocoa. Species within the genus *Pseudococcidae* are the only carriers of the cocoa swollen-shoot virus and are therefore of great importance. *Planococcus citri* is found in all cocoa areas. *P. dilacimus* is found from Sri Lanka to Papua New Guinea. *P. njalensis* is the major carrier of swollen-shoot virus and is distributed throughout West Africa, where *Planococcus hargreavesii* and *Planococcus kenyae* are also found. *Ferrisiana* (*Ferrisia*) *virgata* is of minor importance throughout the tropics. Adult females and eggs of some species are covered in powdery white wax.

Scale insects and mealybugs are usually attended by ants of a variety of species, which often construct tents over these insects.

Control is difficult because of the covering of wax and the water-repellent skin. Biological control has had no effect so far. Control of ants by insecticides reduced numbers significantly, with some reduction of disease, but allowed other pest insects to multiply. Systemic pesticides reduced the incidence of disease for a short period only; this was not economic. Removal of alternative virus hosts, particularly *Cola chlamydantha*, *Cola gigantea*, *Ceiba pentandra* (kapok), *Bombax buonopozense* and *Sterculia tragacantha*, helps to minimize the incidence of diseases.

For a fuller discussion, with references, see Wood and Lass (1985c).

Thrips

The species most frequently found on cocoa is *Selenothrips rubrocinctus*, the cocoa or red-banded thrips. This is a small insect, pale yellow with a red band when young and black when mature.

Leaves become silvered following sap-sucking. Eggs are laid on the underside of the leaves. The nymphs drop fluid on the leaves, which forms brown spots. These spots on dry or silvered leaves are characteristic of thrips.

It has been shown (Fennah, 1965) that the presence of thrips indicates the presence of another problem, such as nutritional imbalance, poor soil conditions and sudden change of shade level. Thrips can be controlled by spraying BHC, fenitrothion, diazinon or propoxur, but the long-term cure is to alleviate the adverse conditions that encourage thrips.

Ring Bark Borers

Endoclyta (or *Phassus*) *hosei* and *Phassus sericeus* (or *damur*) destroy the bark in a ring around the stem in Sabah and Java, respectively. The ring is covered by woven silk and bark; the larvae leave by a tunnel through the stem. Attacks are usually on trees between 6 months and 3 years old.

Control is helped by removal of *Trema cannabina*, which is the main host plant. This grows profusely after land clearance. The main control is by injecting dieldrin emulsion, where permitted, into holes and sealing with earth. This must be done regularly, following an inspection of trees, at monthly intervals.

Zeuzera coffeae is a pest in Sri Lanka, Malaysia, Java and Papua New Guinea. *Zeuzera roricyanea* is present in Sabah. Larvae tunnel along slender stems for up to 30 cm and then make a transverse tunnel before pupating to become the adult leopard moth. Young seedlings can be killed; affected branches on larger trees often break off.

Control is mainly by removal of affected branches. Clearance of undergrowth and vegetation between forest and plantation reduces the risk of infestation. Systemic insecticides are sometimes effective.

Cocoa Moth (Cocoa Pod Borer)

Conopomorphs (*Acrocercops*) *crameralla* is a small moth which does much damage to pods in Java and the Philippines. It also occurs in Papua New Guinea and the Celebes. Eggs are laid in the epidermis of pods, usually in furrows. The larvae go through the husk and bore around the beans for 15–18 days before leaving to form a cocoon. Damage cannot be seen until the pod is opened and found to be full of frass; the beans are useless.

Some degree of biological control has been achieved by release of *Trichogrammoidea bactroefumata*, an egg parasite (Lim and Chong, 1986).

Alternative host species must be kept away from cocoa, i.e. *Nephelium* spp. (rambutan, litchi, pulasan), *Cola* spp. and *Cynometra cauliflora*. The main control system is to remove all pods from a plantation at the end of a main crop period and to destroy them. This breaks the life cycle of the pests. This system is known as 'Rampassen'. Protection of pods by sealing in polythene bags is effective (Wardojo, 1984). A variety of chemical insecticides have been found to reduce pod-borer infestations significantly. Unfortunately, the best were dieldrin and aldrin. Lim (1992) reviewed the incidence, biology and control of pod borer.

Bollworm

The spiny bollworm of cotton, *Earias biplaga*, attacks cocoa from Côte d'Ivoire to the Congo. It mainly attacks unshaded plants up to 3 years old. It is therefore a greater problem on cocoa planted on clear-felled land, particularly when the early shade is inadequate. Early removal of temporary shade has been followed by severe infestations.

Young larvae feed on apical buds and older ones on leaves. Continuous disbudding can therefore occur, inhibiting the formation of trees with the desired shape and a good canopy.

Control is difficult. The insects thrive, despite the presence of many parasites. A systemic insecticide, such as monocrotophos, is recommended in some areas.

Armyworm

Tiracola plagiata attacks cocoa in Papua New Guinea. Local weed species, particularly *Erecthitis hieraclifolia* and *Euphorbia cyathophora*, which spread on newly clear-felled land, become heavily infested. The pest also lives on *Leucaena glauca* and *Crotalaria anagyroides*, used as shade. The insects feed on the flush and growing-points of the cocoa. Death of cocoa trees is unusual, but canopy development is hindered and trees take longer to come into bearing.

The pest can be controlled by frequent spraying with carbaryl, but this may not be economic. Reduction in *Leucaena* shade might reduce the frequency and cost of spraying.

Leaf-cutting ants

Known only in the American cocoa areas, *Atta cephalotes* is the most important species. It is known as 'suava de mata' in Brazil, 'zampopas' in Costa Rica and 'bachac ants' in Trinidad. The latter term also includes the ant *Acromyrmex octospinosus*, which is an important pest in Trinidad.

Atta spp. form enormous subterranean nests, covering up to 0.25 ha. There they cultivate a fungus, on which they feed. The fungus lives on the pieces of leaves brought in by the ants. *Acromyrmex* spp. form much smaller nests (up to 1 m). These ants cause enormous losses, probably equivalent to losses from locusts in Africa.

Nests are destroyed by introducing heptachlor into the entrances. Infested soil is often fumigated with methyl bromide or phosphine. In Costa Rica, physical destruction of nests of *A. cephalotes* and *Atta colombica*, followed by application of chlordane, is a legal requirement. To control ants in adjacent forest areas, baiting with citrus meal containing mirex is effective.

Ants Living Off Sap-sucking Insects

Azteca spp. in America live off honeydew from aphids, scale insects, hoppers, etc. Some make nests on aerial parts of trees; others use hollow places. The 'enxorto ant' in Brazil, *Azteca paraensis* var. *bondari*, builds a spherical nest, which attracts epiphytes, such as orchids. It feeds on honeydew but damages terminal shoots to get mucilage for nest-building. Terminal defoliation follows and many less vigorous shoots grow to form a broom, below which there is a characteristic cluster of leaves.

Azteca charifex, balata ants, in Trinidad and its var. *spiriti*, the cacarema ant, in Brazil make pendulous nests on sloping stems. Trees with nests do not do well; their pods are dwarfed and scarred with brown patches.

These ants occur very frequently, up to 150 nests ha^{-1}. Control is therefore a major problem and expense, and is achieved by injecting BHC dust into the nests.

Chafer Beetles

Apogonia, *Anomala* and *Chactadoretus* species in Malaysia feed mainly on roots but also damage soft leaves and flowers. Rose beetle, *Adoretus versutus*, defoliates cocoa in Java, Fiji and Samoa. Young plants must be protected by split bamboo, palm leaves or plastic cylinders. Older plants suffer less damage. Treatment of soil or leaves with chlorinated hydrocarbon or systemic insecticide (e.g. phorate) will control these pests.

Camenta obesa feeds on the main roots of cocoa. Young trees (up to 2 years old) can be killed by one or two larvae. They can be controlled by pouring coal-tar derivatives or BHC into holes directed towards the top roots or by granular systemic insecticide, e.g. phorate, on the ground around the collar.

Cocoa Beetle

The longhorn beetle, *Steirastoma breve*, is known from Florida to Argentina in America. It is a pest of cocoa in many countries but not in Brazil or Central America. Eggs are laid in holes in the bark. The larvae bore a chamber in the cambium and bark. From the chamber a tunnel is bored in a spiral, which often rings the stem so that it dies. A pupal chamber is then bored in the heartwood, weakening the stem. A gummy, gelatinous exudate appears around holes through the bark made by larvae. Trees from 6 months to 5 years are attacked. The intensity of attack increases as the amount of shade is reduced.

At one time larvae were dug out from stems, but this weakens the stems. A favoured alternative host plant, *Pachira insignis* ('chataigne maron' in Trinidad), is often cut and placed close to cocoa for 3 weeks in wet or 2 weeks in dry conditions. It is then removed and burnt, being replaced with fresh material. Organochlorine pesticides will kill this species, but their use reduces the number of predators on other pest species so that these pests multiply. Lead arsenate has been used. Several systemic insecticides will kill this pest.

Longhorn Beetles

Beetles of *Glenea* spp. are pests in Papua New Guinea, Java and Malaysia. *Glenea alvensis* and *Glenea lefebueri* prefer heavy shade so are a greater problem in heavily shaded and overgrown plantations. Attacks are much fewer in well-maintained plantations. *Glenea novemguttata* is a pest in Java close to undisturbed forest. Larvae burrow below the bark, ejecting frass through the holes, for up to 3 months and then dig a pupal chamber. The long gallery, up to 20 cm, and chamber weaken stems, which may die. Trees 3 years old and over are susceptible. Control measures are similar to those for *Steirastoma*, with the exception of trapping.

Pantorhytes spp.

At least six species of these weevils attack cocoa. These species are only known in Papua New Guinea, Solomon Islands and adjacent land areas. Larvae burrow in stems, which are therefore weakened. Adults feed on young leaves, the bark of young shoots and pod husk, on which they leave scars up to 1.5 cm across.

Anoplolepis longipes (crazy ants) attack *Pantorhytes* weevils and eat their eggs. O'Donohue (1992) reports several instances where they have eliminated a *Pantorhytes* infestation. Trichlorphon applied by mist-blower kills adults; the treatment must be repeated, as larvae in the trees will continue to mature. Grease-banding will help to prevent further infestation. Application of a systemic insecticide kills larvae and adults. Alternative wild host plants must be kept out of plantations and land immediately adjacent. Smith (1981) reviewed control measures.

Ambrosia Beetles

A large number of species attack cocoa but mostly only when trees are weakened by drought, fungal attack, mechanical damage or old age. In West Africa, *Xylosandrus compactus* (or *Xyleborus morstatti*), more usually found in coffee, attack cocoa seedlings. *Xyleborus ferrugineus* in America is associated with *Ceratocystis* wilt disease, caused by the fungus *Ceratocystis fimbriata*. Control involves keeping plantations clean of wild host plants and removal of dead branches. Benzene hexachloride, endosulfan or a systemic insecticide will keep the insects under control.

Nematodes

Campos *et al.* (1990b) discussed the nematodes that attack cocoa. *Meloidogyne* spp. are widespread in cocoa. They cause stunting of young plants, with yellowing and browning of leaves, which fall before the plant dies. Leaves on trees in the field dry up in a similar manner and stems die back. New stems may grow from the roots in the growing season but the whole plant will die eventually. The main weapon against these pests is to keep nurseries and fields clear of them. Nursery soil should be sterilized before use. If nematodes should invade a field planted with cocoa, several contact or systemic insecticides can get rid of them effectively, enabling the trees to produce a higher yield, as shown by Sosamma *et al.* (1980). Shade or other trees planted in or near cocoa should not encourage these pests.

Termites

These do not normally attack healthy trees. Occasional attacks following damage or drought stress can cause severe damage. They can be troublesome in nurseries; incorporation of an insecticide in potting soil will protect the young plants.

Snails

Occasional damage is widespread. In some Pacific islands, the giant African snail, *Achatina fulica*, can destroy young seedlings. Alternative host plants, e.g. *Erythrina* shade, must be removed. A low screen will protect nurseries.

Vertebrates

A wide range of small and large mammals are attracted primarily by cocoa pods. Some birds also attack pods. Clearance of a strip of land between the cocoa and forest can minimize attacks. Smaller animals can be attracted and killed by poisoned baits. Some loss is inevitable. Warren and Emamome (1993) reported attempts to control the neotropical red squirrel, which prefers to attack ripe pods.

DISEASES

Cramer (1967b) estimated that diseases caused losses of potential crop amounting to 43.2% in America, 9.0% in Asia, 20.0% in Africa and 13.2% in Oceania.

Cocoa Swollen-shoot Virus

This is a major problem in Ghana and Nigeria. It has also been identified in Côte d'Ivoire, but there it has no serious effects and no measures are taken to control it. Similar symptoms have been reported from other parts of the world; but the effects are minor and no attempts are made to control these diseases. In some countries, the presence of the virus has not been proved.

There are many strains of the virus, which vary in the severity of their effects, transmission vectors and host range. The most virulent (1A or Near Juaben) can kill Amelonado seedlings in a few months and mature trees in 2 years.

The virus produces swellings on chupon and jorquette shoots and roots. Young flush leaves develop red bands, followed by chlorosis close to the veins. Later, a fern-leaf pattern appears and trees look generally yellow. Pods become mottled, smoother than normal and rounded in shape, with fewer beans inside. Infected trees may not show symptoms for up to 2 years.

The virus is transmitted by mealybugs, particularly *P. njalensis* and *P. citri* (Posnette, 1940). Young mealybugs are most important in spreading the virus. The mealybugs acquire the virus within 1.5 h of feeding on infected material and remain infective for 48 h. They are moved by crawling, by air currents and by the ants that feed them. The mealybugs are also found on many other plants in the forest, some of which are also hosts of the swollen shoot. The most important species are *Cola chlamydantha*, *C. gigantea*, *C. pentandra*, *Bombax buonopozense* and *Sterculia tragacantha*.

There is no cure for a virus infection. Control can only be by removal of infected plants, and of adjacent trees where farmers will permit it. Infected trees are cut to below ground level in Ghana. Outbreaks are inspected regularly and other infected trees removed when found. Inspection continues until no infected trees are found in the subsequent 2-year period. Removal of trees within 30 m of an infection was carried out at one time in Nigeria, where the disease is less virulent than in Ghana, but this has now been completely abandoned. Ollennu and Owusu (1988) have shown that removal of infected trees, followed by replanting, does not of itself prevent early reinfection. This could be significantly reduced by maintaining an unplanted band around each cocoa tree. Hughes and Ollennu (1994) discussed the possibility of restricting the effect of the disease by inoculating trees with a mild strain. Owusu *et al.* (1995) reported that this technique achieved considerable success.

In the long term, development of resistant varieties should reduce losses from the virus. Amelonado trees are all susceptible; Amazon trees vary in susceptibility. Control of mealybugs minimizes the spread of the disease. Some trees are less attractive to mealybugs than others.

Cocoa Necrosis Virus

This has been found in Nigeria and Ghana. It is not transmitted by mealybugs. Distinctive leaf symptoms appear, consisting of translucent distorted patches along veins. Shoots wilt and die back but usually recover. This virus has killed seedlings.

Cocoa Mottle-leaf Virus

Also found in Nigeria and Ghana, this disease produces a red mottle on flush leaves. Translucent veins and bands develop on leaves.

Black Pod Disease (Fig. 13.3)

This is caused by the fungi *Phytophthora palmivora*, *P. megakarya*, *P. capsici* and related species, which are present in all growing areas, although one of the three is more active than the other two in any specific situation (Brasier and Griffin, 1979). The incidence of the diseases varies greatly from one region to another. *P. megakarya*, which is present in some African countries, causes a particularly severe form of the disease.

The fungus infects flower cushions, shoots, leaves, seedlings and roots. On some varieties, cankers are also formed. Root infection is an important part of the annual cycle of the fungus (Gregory *et al.*, 1984). Ground cover preventing spores reaching the soil (e.g. cover crop, mulch) could help to disrupt the normal cycle of the disease.

Infected pods develop a translucent spot about 2 days after infection. This develops into a brown patch, which spreads over the whole of the pod surface and turns black. Sporulation over the infected area produces a white or yellow, dusty appearance to the black surface. There is a characteristic fishy smell. The beans become infected after about 15 days and become valueless unless detached from the husk, as happens when they are almost mature.

Cankers arise from infected pods and a pink-red discoloration develops below the diseased bark. On resistant varieties, the canker does not spread and is destroyed by saprophytic fungi. On susceptible varieties, the canker remains for up to 2 years as a source of infection.

Leaf infections develop from the tips and a wet rot of the veins spreads downwards. This often spreads into the stems, which then die back from the tips. This condition is not usually serious in older plants but can kill a high proportion of plants in a nursery.

The disease spreads more quickly in conditions of high humidity. Careful pruning and reduction of shade will minimize the spread of the disease. Removal of material that is sporulating minimizes the risks of further infection.

Fig. 13.3. Cocoa black pod disease (photograph from J.M. Waller, CABI Bioscience).

This is not completely practicable, as infected pods sporulate so soon after infection. All harvested pods should be taken well away from the plantation before opening and opened pods must not be returned.

Traditional chemical control has been by spraying copper compounds; a wide range of compounds and concentrations have been used. Spraying must start before the disease builds up, so as to cover uninfected pods with a protective layer of fungicide.

Persad (1988) reported that metalaxyl reduced the proportion of black pod infection to a level significantly lower than that achieved by copper hydroxide. A mixture of metalaxyl and copper oxychloride (Ridomil) is now used commercially and has been found to be very effective.

The epiphytic bacterium *Pseudomonas fluorescens* from the surface of healthy pods has been found to be more effective than copper oxide or chlorothalonil in controlling black pod disease (Galindo, 1992a).

Konam *et al.* (1995) reported several factors and treatments leading to integrated management of this disease.

Monilia Pod Rot or *Moniliophthora* Pod Rot

This disease is known also as water pod rot or 'Quevedo disease', and is caused by the fungus *Moniliophthora roreri*, formerly known as *Monilia roreri*. It is believed to have appeared originally in the Quevedo area of Ecuador and has since spread to Peru and to Colombia (outside the Amazon basin). It is also known in parts of Venezuela and Panama.

Young pods are infected; older pods seem to be immune. The infection develops internally, producing a lot of liquid. Externally, dark brown spots appear about 1 month after infection and spread to cover the whole pod. They later become covered with white sporulating mycelium. Beans are wholly or partially destroyed.

Reduction of shade by pruning shade trees and cocoa helps to reduce infection. Infected pods should be removed weekly and destroyed. Copper fungicides should be applied weekly at the peak flowering period and thereafter every 10–12 days. Humidity within the canopy should be kept as low as possible by pruning, drainage and shade reduction. Variations in susceptibility of different varieties suggest that there is scope for breeding for resistance.

Jiménez *et al.* (1988) reported that both chlorothalonil and biological control by the epiphytic bacterium *Pseudomonas aureoginosa* significantly reduced the incidence of this disease. Galindo (1992b) has reviewed the importance, epidemiology and control of this disease.

Witches'-broom Disease (Fig. 13.4)

This disease is caused by the fungus *Marasmius perniciosus*, which originated in the Amazon basin and is now found in most South American countries and some West Indian islands.

The major symptoms of the disease are the brooms, which grow from infection of growing buds. They are thicker than a normal shoot and produce many short lateral shoots. Initially green, the brooms turn brown when the host stem dies back. Cushions can be infected; the flowers have abnormally thick stalks. Pods may be malformed and die young. The effects on pods infected after formation depend on their age when infected. Younger pods become distorted and die. On older pods, black speckles appear externally and beans will die if the pod is not harvested promptly.

The fungus is also found in some wild relatives of cocoa in the Amazon area. After infection and development of the symptoms, the fungus is dormant

(a)

(b)

Fig. 13.4. Witches'-broom disease. (a) The broom. (b) Mushrooms on stem. (Photographs from J.M. Waller, CABI Bioscience.)

for several months before producing, on dead brooms and pods, small pink mushrooms, which produce spores. The mushrooms usually appear in wet weather and produce vast numbers of spores, which are distributed by wind. Rudgard *et al.* (1993) reported the epidemiology of this disease.

The most important component of control is regular removal of diseased material. Fifteen centimetres of diseased stem must be removed and also diseased cushions, with some bark, and pods. All such material should be burnt or buried, well away from the cocoa; this clearance should normally be done three times a year. Phytosanitation is most important during the dry season. Andebrhan (1994) reported that there was wide variation between individual trees in resistance to the disease. Identification and removal of susceptible trees substantially reduced the level of infection; in one example, removal of susceptible trees reduced the number of trees, which lost over 50% of their pods from 5–25% of the trees in the field. This shows that there is a substantial proportion of resistant trees, which can be used in development of resistant stock. No economic chemical control has been found, although Laker *et al.* (1988) reported that moncut was very effective in preliminary investigations *in vitro* and *in vivo*. Moncut was found to be effective for eradication of witches'-broom from seeds before the seeds were moved to another country (Ducamp, 1993).

This disease is the subject of an international programme of research, the International Witches'-broom Project (IBWP). Maddison *et al.* (1994) summarized the research carried out and recommendations made on the basis of results achieved to date.

Homewood (1995) reported that a new variety, named *Cacao theobahia*, has been developed in Brazil which is resistant to witches'-broom.

Cushion Gall Disease

This disease is rarely of great severity. It is caused by the fungus *Fusarium rigidiuscula* (or *Fusarium decemcellularae*); its perfect form is *Calonectria rigidiuscula*. Galls of varying types are formed on the infected flower cushions. Trees do not suffer greatly but their yield is reduced, probably because infected cushions cannot produce pods.

Control methods consisted at one time of removal of gall-bearing trees. A systemic fungicide is now recommended. Seed from gall-bearing trees should not be used for propagation.

Mealy Pod Disease

Caused by a fungus allied to *P. palmivora*, this disease does not cause great loss. Entry of the fungus to pods may be restricted to wounds. Infected pods look very similar to those with black rot. Frequent and regular harvesting, along with minimizing wounds to pods (often mainly caused by animals), should prevent heavy losses.

Diplodia Pod Rot and Warty Pod Rot

Botryodiplodia theobromae causes *Diplodia* pod rot, known also as brown pod rot and *Botryodiplodia* pod rot. It only attacks pods with wounds or under severe stress, such as drought. Symptoms are similar to black pod rot but diseased pods are covered with black sooty powder (spores), compared with the white of black pod. Control is by alleviating the causes of stress and avoiding wounds. If an attack is severe, a systemic fungicide, such as iprodione (Rovral), can be applied.

Warty pod rot has spread in recent years in Ghana and Côte d'Ivoire. Pale green protuberances appear on pods. These are invaded by *B. theobromae* when the pods are mature, which starts a soft rot. No pathogenic agent has yet been identified. Caging pods reduces the incidence of warts, so there may be an entomological explanation (Kebe, 1988).

Note that black pod rot (*P. palmivora*) is known in French as 'pourriture brune', while brown pod rot is 'pourriture noire'. The same confusing transcription occurs in Portuguese.

Minor Pod Diseases

Other pod diseases that have been reported as causing minor damage are *Phytophthora hevea* in Malaysia, *Phytophthora megasperma* in Venezuela and *Fusarium roseum* in Costa Rica.

Evans and Prior (1987) reviewed the causal agents and control of cocoa-pod diseases.

Ceratostomella Wilt

The fungus *Ceratocystis fimbriata* enters the trees through damage caused by implements or through holes bored by beetles, particularly *Xyleborus ferrugineus*. It causes severe losses in Colombia, Ecuador and Venezuela, and is known in Costa Rica, Trinidad, Guatemala, Dominican Republic, Hawaii and the Philippines. The disease is known as 'mal de machete'. The fungus is known in Africa but does not attack cocoa.

Whole or part of the tree wilts and the part affected will die rapidly. Mature leaves droop from their normal horizontal position. These leaves remain in position on the dead stem. Insect tunnels are always found in wilted stems. The wood surrounding the wound is always discoloured.

Dieback

Branches die back from the tips. The stems wither initially from the fine twigs backwards; this is followed by browning and drying of the leaves, the youngest first. Leaves usually fall when dry. The condition may affect only a single branch but can spread to kill the whole tree.

Apart from specific fungi, which are dealt with separately, no other specific disease has been proved to cause this condition, although many have been identified on the dead stems. The incidence of dieback may be associated with stress from abnormal climatic conditions, nutrient deficiencies or imbalances or with low levels of shade. Damage from a variety of insects may initiate dieback, even though the insects have been controlled. Capsid attack is a frequent cause and may be aggravated by the fungus *C. rigidiuscula*. Other insects which may induce dieback include *Earias biplaga* in West Africa and *Selenothrips rubrocinctus*, which exists in most cocoa areas.

Control can only be by identifying and alleviating the causes of stress.

Vascular Streak Dieback (Fig. 13.5)

The fungus *O. theobromae* was identified as causing a major problem in many but not all the islands of Papua New Guinea in 1979. A form of dieback known in west Malaysia since 1957 was later found to be the same fungus. It has since been reported in Sabah, Sarawak, the Philippines, southern Thailand, India (Kerala state), Hainan Island (off China) and Indonesia. Attacks are most severe in high-rainfall areas. Amelonado cocoa is more susceptible than Amazon types.

The disease shows first as a yellowing of leaves on the second or third flush. Green spots develop on the yellow background before the leaves fall. The symptoms spread to all the leaves. After leaves have fallen, short shoots grow from leaf axils. Inside diseased stems are bundles of black vascular streaks, which extend further back than the diseased leaves. If unchecked, the disease will kill the tree.

The disease spreads as spores arising from diseased branches. These are released at night under particular climatic conditions and moved by wind. Sunlight kills spores. Dispersal is limited by high humidity and lack of wind.

Selection and planting of resistant material is the main long-term control of this disease. This has been done successfully in Papua New Guinea, where there is a mixture of Amelonado and Trinitario stock. Resistance was found mainly in Trinitario. There is some resistance among the Amelonado, and in some Upper Amazon stock introduced recently. The disease has now been reduced to a minor problem. Some islands in Papua New Guinea are free of the disease and therefore very strict quarantine procedures are in force for inter-island transfers. These procedures have been described by Keane and Prior

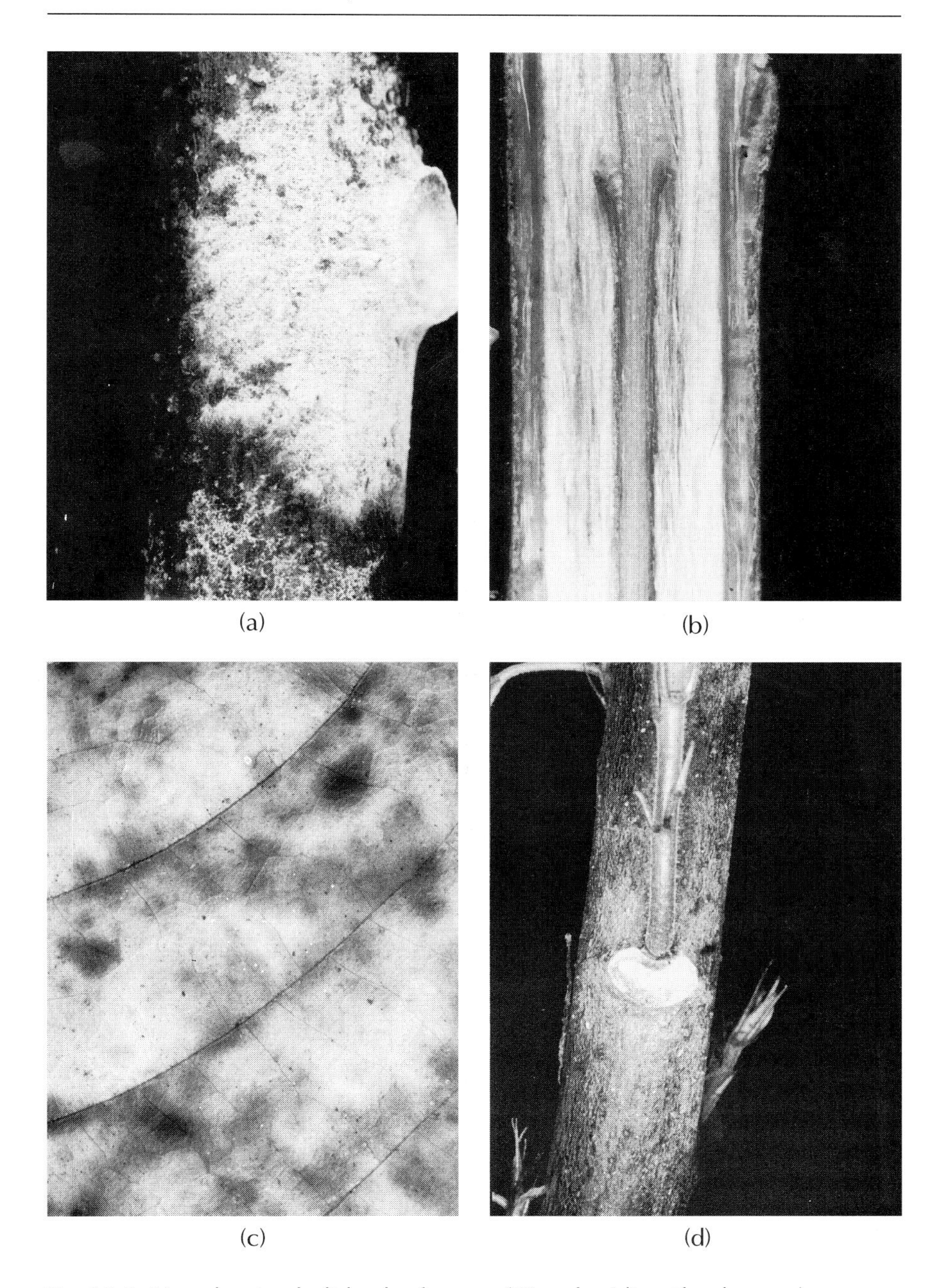

Fig. 13.5. Vascular streak dieback of cocoa (*Oncobasidium theobromae*). (a) Basiodoma; (b) internal streaking; (c) green islands; (d) adventitious shoots. (Photographs from J.N. Hedger, University of Westminster.)

(1992) in a chapter discussing the biology of the disease. Integrated management of the disease in Sabah has been described by Tay *et al.* (1989).

Pruning of diseased branches about 30 cm below diseased xylem minimizes the spread in the plantation. Severed branches need not be removed from the plantation, as the fungus does not sporulate on dead or dying stems. Removal of unwanted branches reduces the number of young leaves that could be infected. Nurseries must be sited well away from diseased cocoa to ensure that plants are free of disease when planted out. It is desirable to raise new plants under a cover, which ensures that leaves of young stock are kept dry except for a few hours after watering.

Sudden-death Disease

This disease, known also as *Verticillium* wilt, is caused by the fungus *Verticillium dahliae*. It occurs in all countries and causes significant damage in Uganda and Brazil.

Leaves wilt suddenly, drooping without loss of colour. They dry off from tips and margins and start to roll up from the margins. Fine branches break off and leaves fall, leaving a bare stem. Trees can die within a week. Internally, petioles, stems and roots discolour, light brown becoming darker, usually in streaks.

The disease is much less severe in cocoa under shade than in unshaded conditions. No chemical control has yet been found. Selection of resistant material may minimize the effects in the long term. Pruning of diseased branches restricts the spread of the disease.

Pink Disease

This is caused by the fungus *Corticium salmonicolor*, a fungus that affects many dicotyledonous crops. This disease does not cause much loss in most areas, but it is of greater importances in some Far Eastern countries. Its incidence has been increasing in Brazil.

The first visible sign is the fruiting bodies on the bark, which appear as flat pink patches or pimples. This happens only after the mycelia have spread well into the bark. Defoliation and death of the affected branch follow.

Sporulation occurs at night and is most active in high humidity. Spores are distributed by wind and need moisture for germination. At a low level of shade, cocoa stems will be kept drier, so that conditions for germination of spores will be less favourable. On the other hand, there will be freer movement of air, so that spores will disperse over a larger area.

The disease will be minimized by a level of shade that enables wet stems to dry out quickly and ground conditions that permit speedy drainage of surface water. There are many alternative hosts; the most susceptible should be kept away from the plantations. Some shade and cover crops, particularly *Cajanus cajan* (pigeon pea), *Crotalaria* spp. and *Tephrosia* spp., can therefore increase the incidence of pink disease and should be avoided. Regular spraying may be necessary if losses are heavy; Bordeaux mixture is reported to be effective (Thorold, 1953); several other effective fungicides have been reported from Brazil (Anon., 1983). Pruning of infected material will minimize the spread of the disease.

Luz *et al.* (1985) investigated the relationships between pink disease and climatic factors. They found correlations with low temperatures, low vapour pressure and high relative humidity.

Thread Blights

Marasmius scandens produces white thread blight and *Marasmius equicrinus* horsehair blight. The first kills leaves, which become covered by white mycelial strands. The second produces black fungal threads throughout the canopy. These prevent leaf fall and inhibit the growth of new leaves.

Removal of dead material minimizes further damage. Pruning to reduce humidity in the canopy and removal of infected wood can also help. Spraying of a copper fungicide would kill the fungus but may not be justified, as losses from these diseases are usually small.

Brown Root Diseases

The fungus *Phellinus noxius* (formerly *Fomes noxius*) is widespread in West Africa and the Far East. Although usually of minor importance, it causes significant damage in Papua New Guinea.

Leaves wilt rapidly on the diseased tree, followed quickly by death of the tree. Mycelia on the roots are golden before tree death but turn brown and hold a crust of soil around the roots. Roots and mycelia eventually turn black.

White Root Disease

This is caused by the fungus *Rigidoporus lignosus* (formerly *Fomes lignosus*), which produces white rhizomorphs on the roots, which age to orange-red. Fruiting bodies form on the collar and have an orange-yellow upper surface and orange, red or brown lower surface.

Collar Crack

The causal fungus *Armillaria mellea* (honey fungus) flourishes in wet conditions and is distributed throughout the world. Fruiting bodies, light brown, changing to yellow and then black, appear at the base of the tree. Mycelia invade the tree, causing the wood to split and making the collar crack for up to 2 m in length. The tree often falls over before it wilts.

Black Root Disease

Three *Rosellinia* species have been found on cocoa, but only *Rosellinia pepo* in the West Indies is of any importance. Infected roots are covered with smoky-grey mycelium, which turns purplish black; minute spherical fruit bodies are embedded in it. Trees wilt and leaves die before the final death of the tree.

All root diseases usually arise from infected roots left in the ground from forest trees or former tree crops. Infection of cocoa is usually by contact between roots and infected residues or mycelial inoculum in the soil, although spores may infect through wounds on stems or roots. The infection is often limited to a single tree. Removal of the infected tree, with all its roots and any other roots that are found, will usually prevent the infection from spreading. Removal of adjacent trees will sometimes be necessary. It is desirable to leave the hole at the tree site open for several months; this helps to kill any remaining pathogen inoculum. The hole can be refilled later and a new tree planted. For a fuller discussion see Waller and Holderness (1997).

MISTLETOES

Several species of these parasites are found on cocoa, and their presence has been reported in most growing areas. In West Africa, *Tapinenthus bangwensis* is the most common; this has red flowers and berries. Next in frequency is *Phragmanthera incana*, with yellow flowers and blue fruit; four other species occur in small numbers (Room, 1972). Kuijt (1964) reported problems with *Oryctanthus* spp. and *Phoradendron piperioides* in Costa Rica. Colombia has problems with members of three genera; *Phoradendron*, *Pihivusa* and *Psittacanthus* (Anon., 1980). *Struthanthus dichlortrianthus* is damaging in Trinidad (Briton-Jones, 1934). In Sabah, *Loranthus ferrugineus* and *Dendrophthoe constricta* have become established (Lass, 1985).

As parasites, they extract water and nutrients from the cocoa and kill the branch beyond the zone which they parasitize. They attract wood-borers, whose galleries are invaded by a specific mealybug (*Cataenococcus loranthi*). Ants service the *Cataenococcus* and the mealybugs that carry swollen-shoot virus. Thus mistletoes can indirectly encourage the spread of swollen shoot.

Mistletoes flower twice a year. The sticky berries are eaten by birds, which have to wipe the seeds from their bills on to a suitable branch. Seeds only germinate in unshaded places and can only penetrate the bark of thin branches. The parasites flower first after about 9 months and live for up to 18 years.

Maintenance of top shade minimizes the germination of mistletoe. Those which do germinate should be cut out; it is sufficient to do this annually when the bright berries can be seen. The branch should be cut about 10 mm below the parasite.

A large parasitic plant (*Acanthosyris paulo-alvinii*) damages cocoa in Bahia, Brazil, if not removed when young (Alvim and Seeschauf, 1968).

WEED CONTROL

Cramer (1967b) estimated the annual loss of potential crop due to weeds amounted to 20.7% in America, 8.3% in Asia, 12.0% in Africa and 10.5% in Oceania.

It has been shown that the growth of seedlings in nurseries is reduced significantly by inadequate weed control (Ruinard, 1966), while early crop yields were significantly affected by the presence of weeds (Bonaparte, 1981). Weeds will grow most strongly during the period between land clearance and establishment of complete shading of the ground by the cocoa. Yields from mature cocoa are significantly related to yields when the cocoa is 3–5 years old (Jones and Maliphant, 1958), suggesting that early vigour, which is enhanced by good weed control, influences subsequent tree performance.

Strong growth of weeds can be expected in the hot, humid climate suitable for cocoa. The traditional method of weeding is by cutting the weeds down to about 5 cm above ground level; the cut material is left to form a mulch. This task will probably have to be undertaken monthly, which is a substantial expense. Less frequent weeding may not reduce the overall cost, as the weed foliage will become more dense the longer it grows unchecked. Slashing does not prevent the weeds from growing continuously, thus competing continuously with the cocoa. Cultivation would stop or reduce weed growth for periods but would damage the surface roots of the cocoa.

Control by herbicides therefore offers significant advantages. While the cost of the initial clearance of weed-infested land can be high, the frequency of applications and hence its cost will fall as the weeds are killed. If the soil is not disturbed the reservoir of weed seeds in the topsoil will be reduced, so that regrowth diminishes. Shading by mature cocoa and shade trees is an important factor in weed control in established cocoa.

A number of herbicides have been found to be effective in cocoa without harming the trees. Those which kill foliage on contact are very effective and widely used. Paraquat is commonly used; often a detergent is added to improve its uptake by plants. This destroys all green foliage it contacts. Glyphosate is similar but is systemic and destroys roots also. Care must be taken that no herbicide touches cocoa leaves.

Pre-emergence herbicides prevent regrowth from seeds and can therefore be used on clean soil, such as around young trees recently planted on cleared land. Applied with a contact herbicide, the amount of regrowth will be minimized. Simazine is safe in cocoa but atrazine is harmful (Kasasian and Donelan, 1965). Diuron is more persistent, but on some soils it can affect cocoa. It has been used successfully in nurseries (Snoeck, 1978).

Where a specific type of plant poses a major control problem, herbicides that act only on the type concerned can be used. Dalapon will control grasses, including those with extensive and deep root systems. 2,4-Dichlorophenoxyacetic acid (2,4-D) acts on herbaceous broad-leaf plants, while 2,4,5-trichlorophenoxyethanoic acid (2,4,5-T) acts on woody plants.

Adeyemi (1988) compared the efficacy of herbicide mixtures with that of single herbicides alone. Mixtures were more effective; a suitable combination would control weeds throughout the year with only two applications in the year. In most cocoa-growing areas, advice on the herbicides that are permitted and which are suitable for local conditions is available, with detailed recommendations for their use. Regulations in most countries specify which chemicals may be used. The chemicals listed may be changed from time to time.

14

BOTANY AND PLANT IMPROVEMENT

BOTANY

Tea as a commercial crop includes several species within the genus *Camellia* in the family *Theaceae*. Many other species are valued as ornamental plants because of their flowers: tea has no such value. Tea taxonomy is very complicated, and hybridization early in its commercial history has left an almost total absence of pure plants. Tea grown commercially includes plants which cover a continuous variation of characteristics (Table 14.1). To produce a tea beverage, the essential chemicals must be present in the leaves.

Tea is commonly accepted as *Camellia sinensis* (L) O. Kuntze, irrespective of any variation in characteristics. This is normally a diploid ($2n = 30$ chromosomes) but polyploids occur. Various types within the tea population are often mentioned. Two which are well-known are the China and Assam; less common is the Cambod. These are often referred to as *C. sinensis*, *Camellia assamica* and *C. assamica* subsp. *lasiocalyx*, respectively (Wight, 1962). Alternatively, the China type may be described as *C. sinensis* var. *sinensis* and the Assam type as *C. sinensis* var. *assamica*. The characteristics of each of these taxa have been listed by Mohanan and Sharma (1981).

There are a number of species with similar characteristics which do not produce the necessary chemicals but may be involved in hybridization. Most important are *Camellia irrawadiensis* Barua (Wilson's *Camellia*) and *Camellia taliensis* (Forest's *Camellia*). Commercial tea, therefore, consists of hybrids, which show a range of characteristics from the China type to the Assam type. There is a continuous variation from one type to the other and any population of seedlings will include plants with a range of intermediate characters. The complexity of hybridization in tea, including some introgression of non-tea *Camellia* species, has been discussed by reference to chromosomal complexes by Chaudhuri (1993).

In its natural habitat, tea is an evergreen shrub or small tree (Fig. 14.1). The China type is a dwarf tree, with small, dark green, narrow, largely

Table 14.1. Variability in vegetative characteristics of tea (from Satyanarayana and Sharma, 1986; Banerjee, 1987).

Characteristics	Range of variability
Mean leaf angle (degrees)	50–120
Laminar angle (degrees)	110–125
Internodal length (mm)	15–70
Individual leaf area (mm^2)	120–200
Leaf area index (LAI)	3.5–8.5
Leaf length/breadth ratio	2.0–2.8
Height (cm)	184–539
Girth at collar (cm)	25–42
Branching habit	Acutely orthotropic to plagiotropic
Thickness of branches at 60 cm from ground level (cm)	1.4–4.4
Length of internode between the second and third leaves from the apical bud of flush shoot (cm)	0.9–3.2
Length of the third leaf from the apical bud of growing shoot (cm)	2.0–6.0
Breadth of the third leaf from the apical bud of flush shoot (cm)	1.5–3.8
Angle between the third leaf of flush shoot and the internode above (degrees)	35–65
Colour of mature leaf	Light green to dark green
Pubescence on the bud and abaxial side of the first leaf	Glabrous to densely pubescent
Anthocyanin pigmentation in young leaves and petioles	Nil to dark
Dry weight of flush shoot (three leaves and a bud) (mg)	60–350

serrated erect leaves. The flowers are borne singly. The Assam type is a taller tree, with larger, less serrated, leaves, which form a greater angle with the stem and tend to droop at their outer point. The colour varies; it is usually lighter green than China and is sometimes very light, almost yellow. The flowers are borne in clusters of two to four.

Seedlings form strong tap roots, while cuttings produce up to five main roots. Lateral branches support a surface mat of feeding roots. Roots penetrate very deeply into soil when necessary to reach water and accumulate a high content of starch, a matter of great practical importance.

Seedlings produce a single main stem. Branches grow from leaf axils; single-leaf cuttings therefore produce one stem from the leaf axil. Leaves are alternate, and glossy on the upper surface. Vegetative growth occurs in flushes of four to seven normal leaves above two scale leaves, followed by a period of

Fig. 14.1. A mature tea bush (height restricted by pruning), Kenya (photograph by K. Willson).

dormancy. The dormant state may occur after fewer leaves if the plant is under stress, such as drought or low air temperature. The first scale leaf is small and drops off, while the second is larger, although smaller than normal leaves. Less serrated than normal leaves, it is known as the 'fish leaf' (Fig. 14.2). The bud in the axil of the highest leaf of a flush becomes enlarged and is dormant; this state is known as 'banjhi' (Fig. 14.3). These buds ultimately break to start new flushes. The crop material required for beverage production consists of the actively growing bud with a length of stem carrying the next two leaves ('two and a bud') or the next three leaves ('three and a bud') (Fig. 14.4).

Smith *et al.* (1990) studied the pattern of tea shoot growth and established the forms of the mathematical relationships between shoot length, dry weight, fresh weight and time. The parameters of the various relationships differed from one clone to another, which provides a basis for examining genetic influences on yield. Gail Smith *et al.* (1994) studied the effects of clone and irrigation on the stomatal conductance and photosynthetic rate of tea.

Flowers are up to 4 cm in diameter, white or tinged pink and slightly fragrant (Figs 14.5 and 14.6). They form in scale-leaf axils, either singly or in a cluster, and develop to form three lobed capsules with thick brownish-green walls. These fruits take nearly 12 months to mature to produce one or two almost spherical seeds in each lobe, seed diameter 1–1.5 cm. Seeds

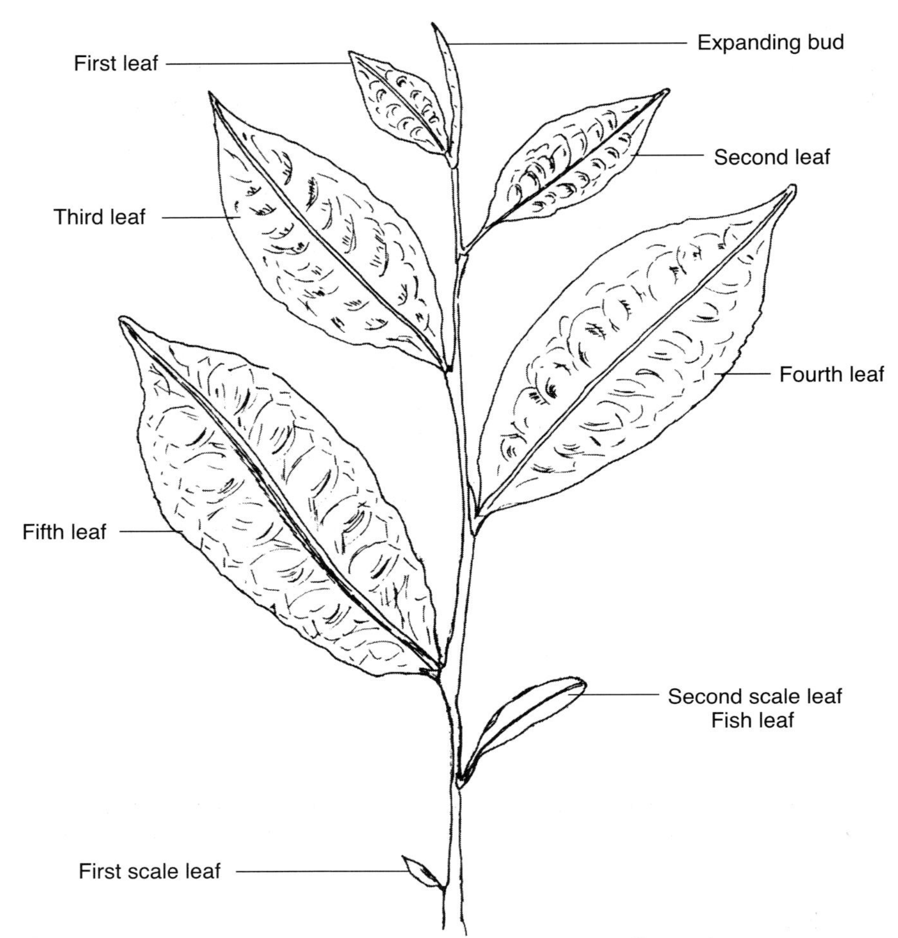

Fig. 14.2. Stem of tea, carrying a flush of leaves which is still growing.

are light-brown with a hard skin and contain cotyledons with a high oil content. Pollination is by insects; most flowers are cross-pollinated. Selfing produces a low yield of inferior seed. Propagation of superior plants in this way is therefore not viable.

Tea grown commercially in China, Japan and their neighbours has developed from the tea indigenous to southern China. It is reasonable to assume that this was the China type *C. sinensis* (var. *sinensis*). The major part of the market in these countries is for 'green' (unfermented) tea, for which the China plant is well suited. Some hybrids with Assam are now used also and there is some production of 'black' (fermented) tea. The industry and the tea plants

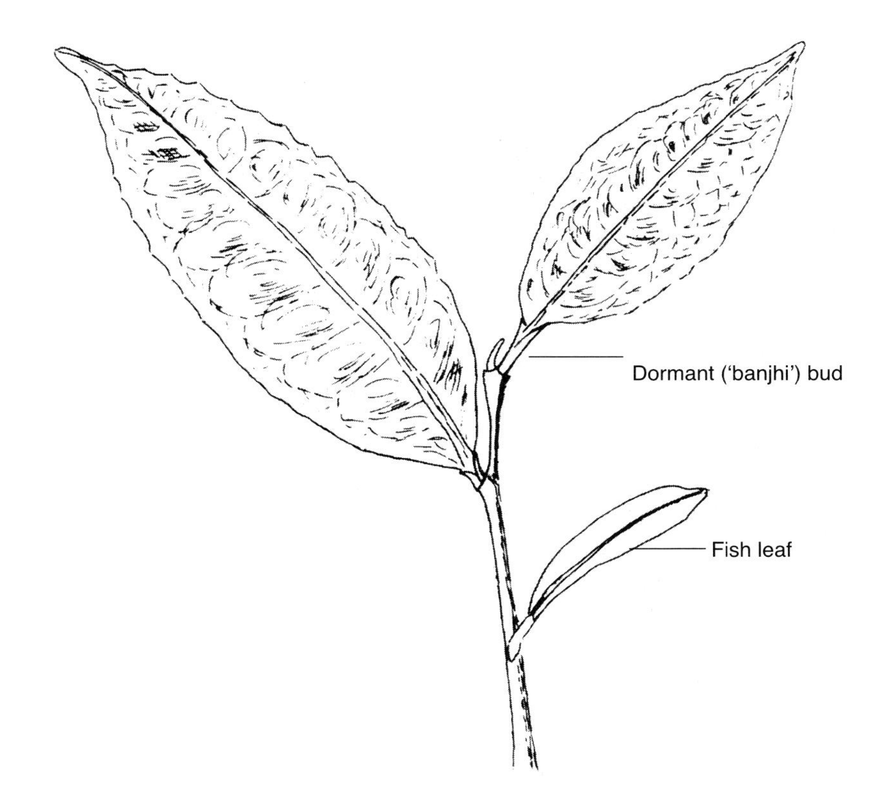

Fig. 14.3. Tea shoot which has stopped growing, as the uppermost bud has become dormant ('banjhi') after only two normal leaves have been formed above the fish leaf.

grown have developed independently of the tea grown elsewhere in the world. The latter originated from the tea planted in Assam, of which the first commercial plants were imported from China. The presence of tea indigenous to Assam had been reported, but there was no knowledge of how to propagate and make tea from it. After the introduction of China tea to Assam, plants of the two types were planted together so that hybridization occurred between the two to give a stock with the characteristics of both types in a range of proportions. Seed orchards were planted, using locally available material, thus establishing hybrid seed as the basis of the developing industry.

The yield of usable crop for a given level of input is dependent on the partition of the dry matter produced by photosynthesis between different plant organs. Significant differences have been reported in partition between different growing areas. Table 14.2 shows that the proportion of total dry-

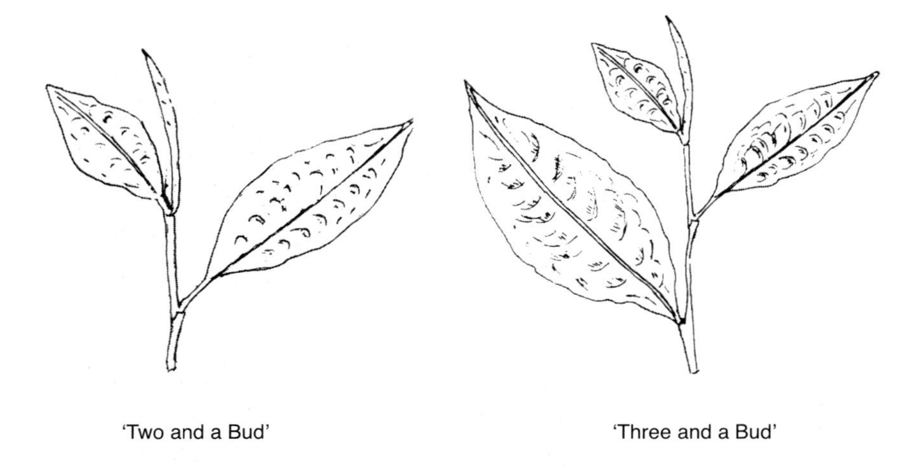

Fig. 14.4. The two normal standards for harvesting ('plucking') of tea leaves from the bushes.

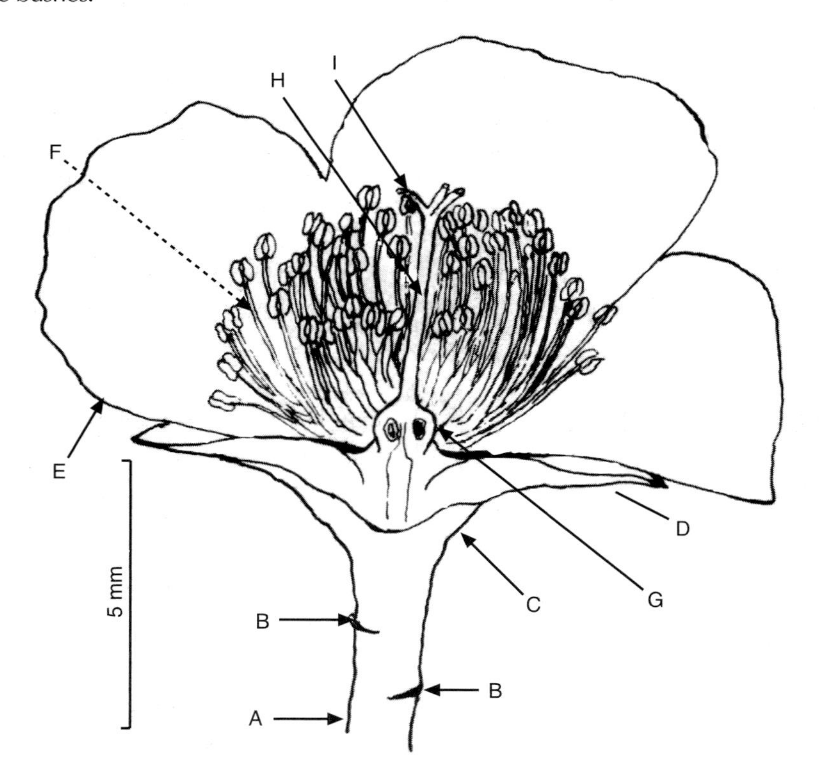

Fig. 14.5. Section through the tea flower. A, pedicel; B, scars; C, receptacle; D, sepal; E, petal; F, stamen; G, ovary; H, column; I, style arms. (From Barua, 1963.)

Fig. 14.6. Tea flower (photograph from Tea Research Association, India).

Table 14.2. Partition of dry matter in tea (from Othieno, 1982; Barua, 1987).

	Percentages of dry matter produced located in various organs		
Location	Crop	Stems and lower leaves	Roots
Assam*	29.4	62.1	8.5
Kenya†	11.3	71.6	17.1

*Mean of 25 clones with a range of leaf-pose categories.
†Mean of five clones.

matter production which becomes harvestable crop is, in Assam, almost three times the proportion achieved in Kenya. There are no reports of detailed studies of the reasons for this, which could be varietal, climatic or some other factor. If varietal, it is a fact to be taken into consideration in plant improvement programmes.

PLANT IMPROVEMENT

The vegetative characteristics of tea plants vary over wide ranges (see Table 14.1). It is therefore not surprising that the yield of leaf from individual plants and the quality of the tea produced vary widely.

The first step in improving overall yield and quality was to select individual plants which gave high yields of high-quality tea and to propagate these vegetatively as clones. This required the development of a method of raising plants from cuttings. Current procedures are described in Chapter 16. The selected material has then to be bulked to produce sufficient plants to give enough leaf for processing for quality assessment.

Clones have been selected from fields of tea in production, from seedling nurseries and from special plantings of selected seeds. Various criteria for selection for yield have been used. The potential of individuals may be indicated by the vigour of their growth, which can be assessed by the number of flushing stems in a given surface area multiplied by the area of the top of the bush. The spread of a bush can, however, be affected by local soil variations and the vigour of adjoining bushes. Specific characters, for example leaf colour, leaf angle, pubescence, size (weight) of two leaves and a bud and leaf area index may be selected where other superior plants have been found to have similar characteristics. None of these criteria are infallible. Comparisons with neighbouring plants are dependent on consistent soil and the quality of the neighbours. Specific characters are rarely directly related to yield.

Quality is equally important in selecting potential clones. There is evidence that pubescence is related to quality (Fig. 14.7) (Wight and Gilchrist, 1959). Potential selections are raised vegetatively and then pruned and allowed to regrow. Where regrowth has reached a reasonable height, the growth can be measured by pruning again and weighing the prunings. Alternatively, the top of the bush can be levelled ready for plucking and the 'tippings' (the foliage removed above the level surface) weighed. Leaves can then be harvested from each bush. By comparison of the weights from individual bushes, those giving the higher yields can be selected and a preliminary estimate of fermentability can be made by the chloroform test (Sanderson, 1963). The phloem index (Wight and Gilchrist, 1959) (the frequency of calcium oxalate crystals in parenchymatous cells of the leaf petiole) is also a useful parameter indicating quality (Fig. 14.8). Chemical analysis to determine the proportion of theaflavins present in the leaf can also be used to assess quality potential (Hilton and Ellis, 1972).

Cuttings are raised from selected plants to enable a small block of each clone to be planted. Some selections may be eliminated on the way because they are difficult to propagate. From a small block, it will be possible to obtain some data on yield, which can be compared with the yield of a seedling block of the same age. The leaf from a small block may be sufficient to make into tea, using miniature equipment. This is used for a preliminary assessment of quality in comparison with seedling plants.

Cuttings must be taken from the initial small block and raised to enable a larger block to be planted. It is essential to have sufficient plants of all clones under test to permit planting of a randomized block experiment with control plots of seedlings. This will enable data to be examined for statistical significance. The original material may need multiplying more than once to provide enough plants.

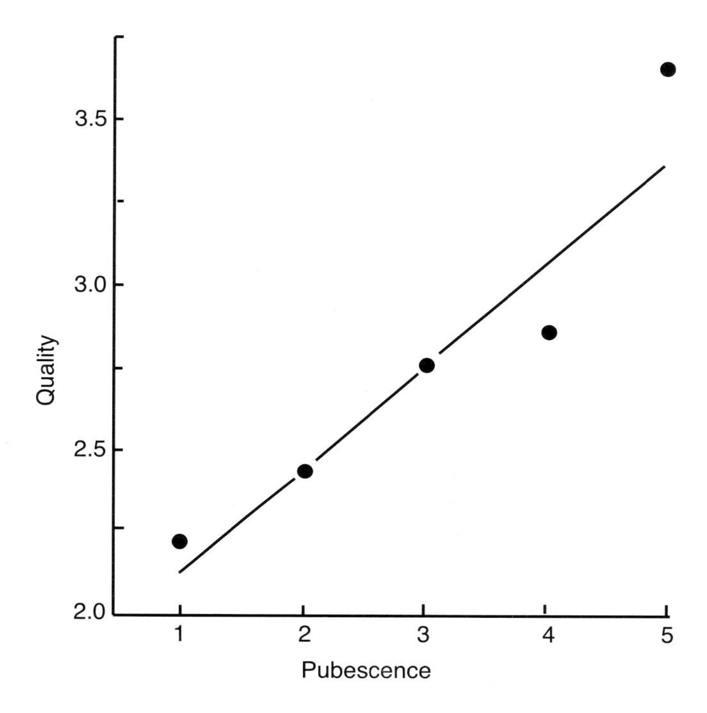

Fig. 14.7. Relationship between pubescence of leaves and quality of tea produced from the leaves. Quality is represented by an arbitrary scale. (From Wight and Gilchrist, 1959.)

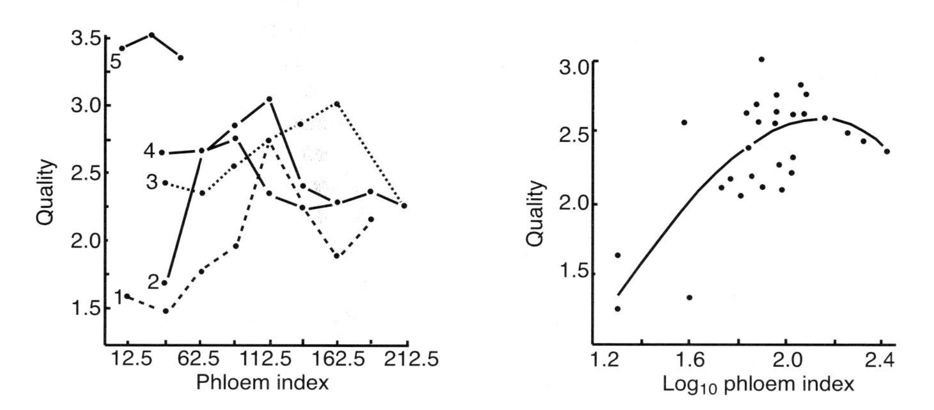

Fig. 14.8. Correlations between phloem index and quality of tea produced. Quality is represented by an arbitrary scale. (From Wight and Gilchrist, 1959.)

Some clones may be found unsuitable during the multiplication process. Records must be kept of important characteristics. In the nursery, the proportion of cuttings that root easily, the growth rate of cuttings, the proportion of cuttings that survive hardening off and the size of plants on transfer to the field are all relevant. In the field, important criteria are the proportion that survive planting out and the degree of branching, which controls the method of shaping the bushes, 'bringing into bearing', for crop production. In due course, the time from planting to tipping at a consistent height and the weight of tippings, which is the first crop, will be recorded. Thereafter, regular harvesting will give comparative yield figures. The weight of prunings at the end of a pruning cycle gives a useful comparison, and the speed of recovery from prune can be important. Harvesting will produce sufficient leaf for a series of miniature processing runs to be carried out. These provide tea for tasting by experts for comparative assessment of quality. These tests need to be repeated at intervals throughout the year.

Resistance to pests and diseases has not been a criterion of great importance so far in tea selection. Relatively severe damage from pest or disease which affected specific clones consistently during the early stages of multiplication would be likely to lead to abandonment of these clones.

The procedure described above will lead, after up to 10 years, to the isolation of a few clones of superior characteristics in terms of yield and quality. It has been estimated that about one in 40,000 seedlings will have superior yield potential. Testing of these selected clones will continue, with comparative trials on larger areas and in other ecological zones. A few clones have become widely accepted; many new plantings in recent years have been clonal – occasionally in monoclonal blocks but more commonly in mixed stands of several clones. Either way, a substantial improvement in yield and quality is guaranteed compared with seedling plantings. The even growth of clonal plantings makes it easier to schedule plucking and other operations, such as pruning, at the optimum time. Processing of tea from monoclonal fields separately from the normal run of manufacture has been carried out.

The potential for yield and quality improvement by selection within specific plantings may be limited by the limited genetic base, as many seedling plantings will have been on seeds from one seed orchard, which is known in the industry by the Assamese word 'bari'. Seed from a specific bari is known as a 'jat'. Attempts have been made to produce improved seed by making a bari from plants selected within a seedling population, using criteria similar to those employed in selecting clones. This approach did not produce any significant advances in yield or quality. Controlled pollination among selected parents may give seeds of higher potential. These have to be grown into plants, among which a further selection can be made for those with significantly better characteristics.

This approach has led to some greatly improved plants. It is a lengthy process. It starts with hand-pollination of selected plants. The progeny have then to be compared in a long-term trial and the best of these are then planted in an isolated bari. The progeny from natural pollination is then evaluated, either as mixed seedlings or as selected clones. From start to release of seed takes about 25 years.

The most successful improvement programme has been to plant selected clones in an isolated orchard and to select clones from the resultant progeny. The selected clones will include improved genotypes; some plants of their progeny should therefore incorporate higher proportions of these genotypes and have superior characteristics. The progeny may come either from controlled cross-pollination of two selected clones or by random fertilization within the orchard. Some biclonal seed has already been released for distribution (Singh, I.D., 1992).

The species of *Camellia* which do not produce tea have usually been left out of programmes of controlled hybridization. In general, such hybrids have produced inferior tea and were not acceptable. There are, however, one or two exceptions among these species which might be worth including in some research programmes. The relatively limited genetic base of much of the world's tea suggests that widening the genetic base could be beneficial in the long run.

There may be scope for introducing non-conventional approaches to tea improvement. Polyploids occur to a limited extent and there are indications that some would have valuable characteristics. Radioactive mutation could produce improved plants. Tissue culture could speed up differentiation and selection. The potential of such methods is limited by lack of knowledge of some aspects of tea genetics.

15

Climatic Limitations, Soil Requirements and Management

Tea developed in the understorey of a dense tropical rain forest south of the Himalayan mountain range and in a less heavily forested environment north of the Himalayas. These give an indication of the conditions natural to the species. These include high rainfall, evenly spread over a large proportion of the year. Temperature would be lower than tropical ambient temperature at low altitude, because of shading in the forest, and humidity would be fairly high. Soil temperature would not be excessive, as the surface would be shaded. Chemically, the soil would be highly leached, due to the high rainfall, and therefore acid, with a low availability of base elements. The China variety, indigenous north of the Himalayas and to higher altitudes, is more resistant to low temperatures than Assam or Cambod types. The conditions necessary for tea can now be defined and understood from the above. The effects of varying some of the conditions can now be quantified.

CLIMATE

Mean rainfall and temperature data for tropical and subtropical tea areas are listed in Table 15.1.

The need for temperature below tropical ambient at low altitude has dictated that, in Equatorial regions, tea is usually planted at high altitudes (1000–3000 m). The optimum altitude drops as the distance from the Equator increases, and tea is grown close to sea level at the extremes of its range – 42°N (Georgia) and 27°S (Argentina).

The optimum air temperature for vegetative growth is 18–30°C. Leaves should be above 21°C. Growth virtually stops at air temperatures below 13°C or over 35°C. There is a linear relationship between shoot extension rate and temperature between 17°C and 25°C (Squire, 1979). The time for growth of individual shoots from initiation to reaching harvestable size, the 'shoot replacement cycle', is related to temperature; its length ('thermal time') is

Table 15.1. Mean annual rainfall and lowest and highest mean month temperatures in tropical and sub-tropical tea growing areas in the world (from Othieno, 1991a).

					Temperatures			
		Latitude	Longitude	Annual rainfall (mm)	Lowest month (°C)	Highest month (°C)	Mean (°C)	Difference between highest and lowest (°C)
Tropical areas								
Kenya:	Kericho	0°22´S	35°21´E	2100	14.5	17.0	15.8	2.5
Uganda:	Rwebitaba	0°39´S	30°26´E	1450	17.0	19.0	18.0	2.0
Tanzania:	Usambaras	5°08´S	38°37´E	1800	17.5	22.5	20.0	5.0
	Mufindi	8°30´S	35°05´E	1950	13.2	18.0	15.6	4.8
Indonesia:	West Java	7°N	110°E	3500	22.2	22.8	22.5	0.6
Sri Lanka:	Kandy	8°N	81°E	2375	22.8	25.8	23.9	3.0
	Nuwara Eliya	7°N	80°45´E	2255	13.3	15.6	14.4	2.3
South India:	Coonoor	11°15´N	76°40´E	2135	15.6	21.1	18.4	5.5
	Annamalais (Cinchona)	10°15´N	76°45´E	3500	–	–	–	5.5
Subtropical areas								
Central Africa:	Malawi (Mulanje)	16°05´S	35°35´E	1590	10.5	30.0	20.0	19.5
Taiwan:	Kilung	24°N	121°E	1933	14.4	28.3	21.1	13.9
North India:	Assam	26°47´N	94°12´E	2085	15.6	28.9	22.1	13.3
	Cachar			1850	18.3	28.9	23.3	10.6
	Darjeeling	26°55´N	88°12´E	2050	6.0	16.7	11.1	11.7
China:	Zhejiang	30°N	120°E	–	0.6	25.0	12.8	24.4
	Yunnan	18°N	105°E	–	2.8	27.2	15.6	24.4
Japan:	Kyoto	35°N	136°E	–	5.0	26.7	15.6	21.7
	Kanaya	34°30´N	138°E	–	5.0	25.6	15.0	20.6
Turkey:	Rize	41°N	40°30´E	2233	6.7	22.3	14.0	15.6
Georgia (Chakwa)		42°15´N	43°30´E	–	–5.6	26.1	15.0	31.7

approximately 475 days °C^{-1} (Tanton, 1982a). The effect of air temperature can be illustrated by the following data for thermal time: Bangladesh, summer (26–29°C) 30 days, winter (16–18°C) 70–80 days, Kenya (15–17°C) 90–115 days. The minimum temperature and shoot extension rate may vary a little from one clone to another. Figure 15.1 demonstrates the relationship between monthly mean air temperature and mean length of shoot replacement cycle.

The absence of shade in many tea plantings leads to high saturation deficits (the difference between the saturated vapour pressure at the given temperature and the actual vapour pressure). The critical value below which shoot growth is inhibited is 2.3 kPa (i.e. relative humidity 28% at 25°C, 45% at 30°C) (Tanton, 1982b). With a high saturation deficit, the normal relationship between shoot growth and air temperature does not hold. Therefore, under such conditions, growth and yield will fall, although there may be adequate water in the soil from rain or irrigation.

Soil temperature is also related to growth, with an optimum range, over which there is a linear relationship, of 19–22°C.

Solar radiation is important. The amount of radiation reaching a tea canopy at high altitude in Equatorial regions can be up to 600 W m^{-2}. Healthy plants absorb most of the incident radiation. Depending on canopy structure, leaf area and orientation, photosynthesis for the entire plant will continue up to high levels of radiation. The remaining energy has to be dissipated as heat. Hence evaporative cooling can be of critical importance, but the stomatal conductance reduces at high saturation deficits, as the stomata narrow, which limits evaporative cooling. Leaf temperature will therefore rise significantly, up to 12°C for every 100 W m^{-2} of radiation.

The canopy structure will affect the tea response to shade, with the Assam types benefiting from shade to a greater extent than the China type, due to their differences in leaf angle, as shown by Hadfield (1968, 1974b).

In low-altitude areas not far from the Equator, high air temperatures and high levels of incident radiation can lead to saturation deficits in excess of 2.3 kPa for long periods of time, with consequent loss of production. In such a situation, some amelioration of conditions at the plucking table is desirable. Hence, shade trees are planted among tea in low-lying areas of India and Bangladesh, whereas they are unnecessary, often disadvantageous, in other areas (Hadfield, 1974a).

The rate of transpiration (E_t) is normally about 85% of the rate of evaporation from an open water surface (E_o). This 'crop factor' of 0.85 is of the same order as that for many mature crops with complete ground cover (Penman, 1948) but may drop to 0.70 in conditions of high saturation deficit (Squire and Callander, 1981). The rate of transpiration may exceed 150 mm $month^{-1}$ in some tea areas, and may exceed 100 mm $month^{-1}$ at times in other areas. Anandacoomaraswamy and Campbell (1993) have developed a model that simulates the water use of mature tea.

The amount of rainfall necessary to ensure that water is available to meet

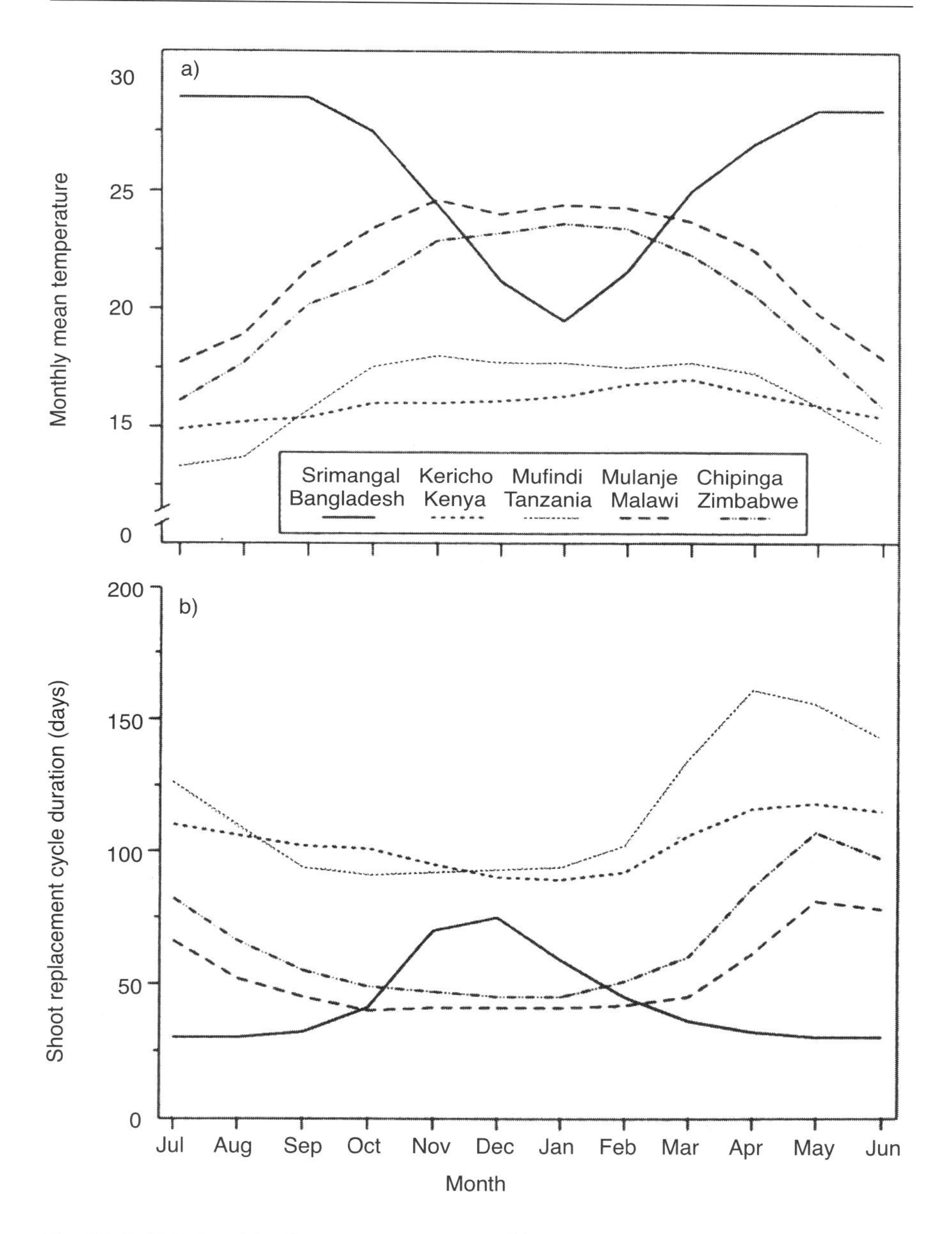

Fig. 15.1. Relationships between mean monthly air temperature and mean duration of shoot-replacement cycle (from Carr and Stephens, 1991, by permission of the authors).

the evaporative demand depends on its distribution and on temperature and other parameters in the dry season. The need for irrigation, the amount of water needed and the optimum times of application must be worked out in relation to climatic parameters for each location. Stephens *et al.* (1994) studied the yield and water-use efficiency of clonal tea.

Some rainfall is lost by runoff, particularly in heavy storms and at the peak of the wet season when the soil is saturated. A full cover of tea and a good mulch on the ground below will minimize runoff and ensure that the maximum amount of water soaks into the ground. Rainfall that is absorbed by soil in excess of its capacity drains away through the profile. A good depth of soil ensures that water which has drained to a lower level can remain within reach of tea. Tea roots have been noted growing to over 15 m depth in suitable soil.

Calculations from actual situations lead to the conclusion that rainfall should average at least 1150 mm for a rain-fed tea plantation; below this level, irrigation will be essential. The distribution of rainfall, climatic conditions in the dry season and the potential water capacity of the soil will have an impact on the average rainfall requirement. For any specific situation, the relevant factors should be determined. Where rainfall is marginal, measures to improve the proportion of the rainfall absorbed by the soil can be valuable. Various erosion-control systems can improve water absorption. Wilkie (1994) reviewed the responses to irrigation and the economics thereof.

The plants can be managed in ways that will improve their water uptake and minimize water loss. Bringing into bearing must ensure the best root development. Mulching of young tea minimizes water loss from bare soil and protects roots from excessive temperatures. It has been shown that mulching is vital for young tea in Malawi (Grice, 1984).

The time of pruning mature tea can be chosen in relation to water demand and availability. Pruning, which usually removes most, if not all, the leaves, reduces transpiration to nil and the prunings form a mulch. If the dry season is cold, pruning can take place soon after the rains end. If this season is warm, crop will increase as temperature rises and pruning will have to be delayed until production falls but should be before the water in the soil has been reduced to a level where it will adversely affect recovery and yield at the start of the next pruning cycle.

The effect of soil temperature on growth varies. Tanton (1982b) found no effect between 18°C and 25°C. Othieno (1979) reported a linear relationship between 15°C and 18°C.

Wind can be a problem in some areas. Wind-breaks across the prevailing wind can be helpful but can reduce crop in adjacent tea due to competition for moisture, shading and loss of a row of tea plants. Benefits are more certain where conditions are extreme, such as seasonal average air temperatures close to the lower optimum level of 18°C, periodical presence of a hot, dry wind or a small area through which wind is funnelled.

Frozen water in various forms can be troublesome. Clear night skies in tropical highlands can reduce air temperatures to zero. Cold air collects in hollows and damages the flush on the tea plants. No permanent damage ensues but some crop is lost. Hail is formed frequently during rainstorms in certain areas and can do a great deal of damage. Leaves are stripped off and bark is severely damaged, a condition which may allow disease to take hold. Heavy loss of crop occurs, averaging at least 10% in some areas, peaking at about 30%. Preventive measures, using rockets, aircraft, etc., have been tried without success (e.g. Henderson, 1966). Snow covers the tea in Georgia each winter, but this protects the plants from excessively low temperatures. The cultivars planted and the management systems take account of the local climate.

There is no evidence that variations in day length in any tea-growing area have any effect on production (Tanton, 1991).

Table 15.1 lists temperatures and rainfall in various tea areas.

SOIL REQUIREMENTS

Tea developed in areas with highly leached soil. Such soils are acid and have low contents of available bases and phosphates, together with large amounts of available iron, manganese and aluminium. Table 15.2 lists some chemical properties of several tea soils.

The highest soil pH shown in Table 15.2 is 5.7, which is above the limit usually quoted – pH up to 5.6. Close to the limit, the amounts of bases available can be critical; in particular, excessive calcium in relation to the other bases. Lime or a high calcium content must be avoided; it is, however, possible to improve some alkaline soils to permit tea planting. On soil of pH 5.7, the calcium level is high; the potassium level is also high so that the soil is just acceptable for tea.

Very acid soils are not disadvantageous but will need addition of phosphate and potassium fertilizers from planting onwards. Other mineral deficiencies may arise, particularly magnesium, zinc, copper and boron. These will need appropriate fertilization as soon as they are identified. Tea accumulates manganese and aluminium from acid soils. Manganese has to reach very high levels in the leaves before it becomes deleterious. Aluminium has no effect which has been reported.

Tea soils have been formed from a wide range of parent rocks: sedimentary from gneiss or granite, flat alluvium, drained peat and volcanic ash are the most common types. All soils should have a high capacity for retention of water but also be free-draining. Tea will grow well on soils of almost any texture, although extreme types may create problems. Heavy clays can be slow to drain and the surface will become waterlogged in heavy rain. Light, sandy soil may have a low water capacity so that tea may run short of water in a drought. Table 15.3 lists physical proportions of some soils on which tea is grown. A good depth is desirable to give a high water-storage capacity.

Table 15.2. General chemical properties of some soils in tea-growing areas.

Country (area)	Soil depth (cm)	pH (soil water 1:2.5)	Organic matter (%)	Total N (%)	C/N (ratio)	Available P (p.p.m.)	Exchangeable bases (mequiv. 100 g^{-1}) K	Mg	Ca	Exchangeable H (mequiv. 100 g^{-1})	CEC (mequiv. 100 g^{-1})	Reference
India												
Assam	0–30 (fallow)	4.73	1.66	0.10	10.05	15	0.96	0.35	0.78	–	5.83	1
Kenya												
Kericho	0–5	5.10	8.46	0.60	8.20	14	2.08	2.50	0.90	1.0	–	
	5–10	4.50	8.00	0.58	8.02	8	1.90	1.40	0.60	3.9	–	1
	10–15	4.60	7.09	0.50	8.24	7	1.74	0.70	0.20	3.8	–	1
	0–23	4.75	3.68	0.10		4	3.08	2.01	0.46	–	20.78	1
Malawi												
Newly planted tea	0–23	5.30	3.00	0.17	10.50	14	1.16	1.31	2.46	–	2.70	1
Replanted tea	0–23	4.15	2.66	0.14	11.46	14	0.69	–	–	–	–	1
Sri Lanka		4.00	5.40	0.33	9.63	24	0.36	–	1.125	–	–	1
Taiwan												
Horizon	I	4.4	1.53	0.09	9.90	1.50	0.15	0.08	1.71	18.50	–	1
	II	4.4	1.38	0.08	10.00	1.00	0.15	0.07	0.41	18.50	–	1
	III	4.6	0.53	0.40	7.80	0	0.16	0.16	0.10	0.53	18.50	1
South India												
Anamallais	0–15	4.9	7.55	0.28	–	22	0.51	1.73	8.83	0.30	11.2	1
	15–90	4.7	2.18	–	–	4	0.45	1.40	4.24	1.05	10.4	1
	90–125	4.6	1.41	–	–	3	0.19	1.48	3.54	4.9	10.4	1
Papua New Guinea	0–15	5.0	–	–	–	13	1.05	1.16	4.8	-	21	2
	0–15	4.3	–	–	–	36	0.74	0.48	2.0	–	24	2

References: 1. Othieno (1991b); 2. K. Willson (unpublished).
N, nitrogen; C, carbon; P, phosphorus; K, potassium; Mg, magnesium; Ca, calcium; H, hydrogen. CEC, cation exchange capacity.

Table 15.3. Physical properties of some tea soils of the world (0–30 cm depth on average) (from Othieno, 1991c).

	A	B	C	D	E	F	G	H	I	J	K	L	M
Texture fractions %													
Coarse sand/gravel	16	–	–	32	21	20	34	24	21	2	2	–	–
Fine sand	34	7	17	18	20	19	33	39	32	4	6	14	49
Silt	27	33	43	28	25	36	22	29	10	11	9	34	30
Clay	17	11	28	5	20	25	10	6	37	82	83	52	1.7
Organic matter	6	33	8	3.4	5.2	–	–	–	4.2	8.5	6.8	2.3	1.7
Texture grade	SaL	SiL	SaL	SiL	SiL	SiL	SaL	SaL	SaL	C	C	CL	CL
Particle density	–	–	–	2.00	2.13	–	–	–	2.61	2.60	2.55	2.71	2.66
Apparent density	–	–	–	1.03	1.07	–	–	–	1.23	0.85	0.77	1.39	1.42
Porosity %	–	–	–	54.3	51.6	–	–	–	53	73	71.3	48.7	46.8
Water-holding capacity %	–	–	–	59.5	65.5	–	–	–	22.5	59.0	60.9	28.2	20.8

A, Bramaputrah alluvium, Assam; B, peat (bheel) soil, Surmah Valley; C, clay flat, Surmah Valley; D, Anamallais, South India; E, Nilgiris, South India; F, Central Province, Sri Lanka; G, Uva, Sri Lanka; H, Pengalengan, Java, Indonesia; I, Usambara, Tanzania; J, Kiambu, Kenya; K, Kericho, Kenya; L, hsing-hua series, Taiwan; M, ho-kang series, Taiwan; –, data not available; SaL, sandy loam; SiL, silty loam; C, clay; CL, clay loam.

Hardpans, rocks and other obstructions that can restrict root growth must be avoided.

Tea soils can be identified by the presence of certain plants among the natural vegetation. These plants are, in general, accumulators of aluminium and often have blue flowers. Many *Albizzia* species have been noted as indicators; others are *Newtonia buchananii*, *Vernonia auriculifera*, *Triumfetta macrophylla*, *Boreria princei*, *Pteridium aquilinum* (bracken), *Dissotis* spp. and *Craterispermum laurinum* (Chenery, 1955).

Organic matter is desirable so long as the soil remains acid. The subsoil of areas cleared for tea often contains little organic matter. The topsoil, which would have a greater organic-matter content, often disappears when the land is cleared. The lack of organic matter should not prevent planting of tea. Good tea management will create an organic top layer under the plants.

Dey (1969) reported a comparison of soils from India and Africa.

SOIL MANAGEMENT

Good management starts with clearing the site. The loss of topsoil must be minimized, which is not easy in high-rainfall areas and particularly on sloping land. Measures designed primarily to minimize erosion reduce the rate of water movement; soil carried by the water will settle out as it slows down. Such measures include wide, shallow terraces across the hill slope with a gentle incline (0.5%), which divert rainwater into a limited number of drains running down the slope (Fig. 15.2). 'Lock-and-spill' drains are rectangular-section ditches running across the slope and falling at about 0.8%. Soil barriers of height lower than the depth of the ditch hold water from which suspended soil will settle out; the speed of movement of water will also be reduced (Fig. 15.3). Microcatchments – disparate lengths of rectangular ditch across the slope – collect water from which suspended soil will settle out. The water soaks into the soil to improve the water reserves in the soil.

Water movement between terraces or drains must be minimized. A suitable cover crop should be planted as soon as possible after levelling. This can be sown broadcast or in lines between the proposed lines of tea. Banks such as the edges of terraces which will not be planted with tea should be planted with a grass (e.g. *Eragrostis curvula*), which will hold the soil but not invade the tea.

The foliage from the original vegetation should be spread as evenly as possible. If the vegetation is light – for example, grass – ploughing and/or rotavating may break it up and distribute it evenly, incorporating it in the soil. Heavier stands up to forest must have the large pieces moved right out of the field, with minimum movement of soil with them. If the smaller pieces cannot be ploughed in, they may have to be burnt. Large fires must be avoided, as the ash concentrates bases to a level that may make the soil alkaline and unsuitable for tea without special treatment.

Fig. 15.2. Sloping field which has been terraced and planted with young tea plants. Note the lines of the terraces sloping at 0.5% across the varying slope of the field, discharging into a cut-off drain, lined with concrete to prevent erosion. This drain runs straight downhill in a gully between the camera and the terraced field, out of sight because of the slope of the ground. Note the straight lines of tea bushes above the terraces. They do not follow the contour, as they were planted to form rectangular experimental plots. A mechanical harvester would require similar straight lines. Kenya. (Photograph by K. Willson.)

Soil which is not sufficiently acid (soil pH above about 5.6) can be acidified. Sulphur mixed with the soil is very effective; some delay between sulphur application and planting of tea is desirable. Quicker results can be obtained using soluble chemicals, such as aluminium sulphate or sulphate of ammonia. However, the latter can, in excess, damage tea plants and its effect is not always permanent.

Tea should preferably be planted in contour lines on sloping land (Fig. 15.4). However, straight lines, as close as possible to across any slope, may be necessary if mechanical harvesting by self-propelled machines is envisaged. Tea lines running up and down slopes encourage erosion, across the slope they minimize it. The cover crop should remain until it is shaded out by the tea, and residual cover plants will be removed when the tea interrows are weeded. The cover crop will start forming a mulch between tea rows. This will be augmented by leaf fall from the tea and ultimately by tea prunings. In some areas, it is beneficial to carry mulch into new tea plantings; the growth rate of the young tea should be improved. Mulching is considered essential in Malawi (Grice, 1984).

Fig. 15.3. 'Lock-and-spill' drains across a hillside in Sri Lanka. The soil is being rehabilitated for tea; Guatemala grass, *Tripsacum laxum*, has been planted in contour lines. (Photograph by K. Willson.)

Fig. 15.4. Tea planted on the contour on sloping land in India (photograph from Dr W. Hadfield, George Williamson & Co. Ltd).

The mulch between tea plants must be allowed to develop. When tea is pruned, the material cut off must be allowed to remain in the tea rows, where it rots down to form a deep mulch layer. The rate of decomposition of this mulch is reduced by the shading from a complete cover of tea. The mulch forms a topsoil of high organic content and nutrient balance appropriate for tea. This is filled with a dense growth of tea roots, which can be exposed by lifting the upper layer of mulch (Fig. 15.5); most of the absorption of mineral nutrients occurs in this layer. Provided that prunings are not removed from the field, a self-perpetuating organic topsoil is formed, where nutrients are freely available. Chemical analysis of soil can give some indication of potential mineral deficiencies, although it is not possible to determine specific critical levels of nutrient availability in soils below which a deficiency will occur or above which a toxicity problem will occur.

A study of Table 15.2 shows a range of low values for total nitrogen. These are not directly related to the organic-matter content. Little nitrogen is needed by a young tea plant. Good management will start applying nitrogen fertilizer as soon as each plant has a significant area of leaf. Experiments have shown that mixing nitrogen fertilizer with the soil at the time of planting often has a deleterious effect on tea plants (O'Shea, 1964).

The available phosphorus is extremely low in most of the soils listed in Table 15.2. Experiments have shown that substantial improvements in the

Fig. 15.5. Tea roots exposed by lifting the upper layer of the mulch, formed by leaf fall and prunings (photograph by K. Willson).

growth of young plants follow mixing a phosphate fertilizer with the soil before planting (O'Shea, 1964).

Some levels of exchangeable potassium are low, while some are much higher. Where levels are low, a small quantity of potassium fertilizer can be mixed with the soil before planting and will counter the tendency for young plants to suffer from potassium deficiency. Potassium does not have to be the base with the highest concentration of the three, potassium, calcium and magnesium. Often it is the one present in the lowest concentration but, if it forms less than about 10% of the total of the bases, there may be a tendency for potassium deficiency to affect growth.

High levels of calcium can induce potassium deficiency and may be associated with a high pH level in the soil. Nevertheless, calcium is often present in higher concentration than potassium, which is usually tolerable provided that the acidity is suitable and calcium is not more than about 80% of the total of the bases.

Levels of magnesium vary widely, but the tea plant is able to absorb sufficient magnesium from a wide variety of soils. Magnesium deficiency is uncommon and is often associated with potassium deficiency. Correcting the potassium deficiency usually corrects the magnesium deficiency also.

Some soils are short of sulphur and deficiency symptoms will appear on young plants unless a sulphur compound can be mixed with the soil before planting. Chu (1975) reviewed the properties of tea soils and their grading for improvement.

REPLANTING OF OLD TEA AREAS AND SOIL REHABILITATION

Rehabilitation and replanting are being carried out on an increasing scale as the availability of clones with ever higher yield potential increases. The soil which has been planted with tea for many years is usually denuded of nutrients and requires rehabilitation. A crop such as Guatemala grass (*Tripsacum laxum*) or a legume, e.g. *Desmodium* spp., is grown for a period up to 3 years (Fig. 15.3). This crop is then ploughed in and the cover crop into which the tea will be planted is sown. The efficacy of various methods of removal of old tea plants and soil rehabilitation has been investigated in several countries, e.g. Kenya (Ng'etich and Othieno, 1993) and South India (Ponnapa, 1991).

Tabageri *et al.* (1991) reported rehabilitating tea without replanting.

Njuguna (1993) investigated the suitability of various clones for replanting and the effects of soil rehabilitation with Guatemala grass on the yield of replanted tea. He reported that, by over 9 years later, the cumulative yield from clonal plants which had been planted into unrehabilitated soil was not significantly different from that of clonal plants which had been planted after soil rehabilitation for 1 year. Clonal plants planted after 2 years' soil rehabilitation had yielded less crop by the same date. A comparison between clones planted into unrehabilitated soil and old seedling tea showed that the clones had out-yielded the seedling tea within 5 years. The replacement of old seedling tea by clones is therefore advantageous, but the advantage of soil rehabilitation is less marked.

16

Field Management

SITE SELECTION

When selecting a site for tea, the climate is the first consideration. The effects on the growth of tea of various climatic parameters have been discussed in Chapter 15. If sufficient data, over a period of years, are available, it may be possible to calculate the potential yield from a particular site, while drainage and irrigation requirements can be predicted (Singh, 1978). Willson (1965) discussed the design and control of irrigation systems. Problems may arise from waterlogging, even flooding, of otherwise suitable sites, particularly if there are periods of relatively heavy rainfall. It is often difficult to forecast such situations. Remote sensing can indicate groundwater reserves and post-monsoon flooding prior to clearance and planting (Barbora and Bordoloi, 1993).

Temperatures should be in the optimum range for as long as possible in a year. Total rainfall and its distribution in relation to temperature distribution will indicate whether irrigation could be necessary. Soil water-holding capacity and depth should suggest how tea will react to the dry season, and how it should be managed.

Data for wind and hailstorms can influence the plantation layout and forecasts of crop yields. Henderson (1966) reported an investigation into hail suppression.

Soil should have acidity (pH) within the desirable range or be capable of the necessary improvement. Concentrations of the various nutrients will indicate which fertilizers may have to be applied.

Physically, the soil should have suitable characteristics and sufficient depth. Areas not suitable for tea may be good enough for fuel wood. The extent to which sloping land can be planted should be assessed and the pattern of drainage – terraces, cut-off drains and streams – worked out. The area required for buildings – factory, workshop, housing and amenities – should be determined and suitable sites identified. Last but not least, the pattern of roadways should be designed to ensure that transport of crop, materials and personnel will be easy throughout the year. Morris (1997) discussed the design, installation and management of estate road systems.

LAND PREPARATION

Existing vegetation, frequently heavy forest (Fig. 16.1), should be cleared, bearing in mind the principles discussed in Chapter 15. Erosion-control measures should be constructed as soon as possible, preferably in the dry season, and the cover crop established before heavy rain can cause severe damage. The lines of tea plants should be marked by a stake at each plant site. Lines of tea should preferably be on the contour on sloping land (see Fig. 15.4, p. 188), but, if harvesting is to be mechanized, it may be necessary to plant in

Fig. 16.1. Interface between a young tea plantation and the indigenous heavy forest, Usambara mountains, Tanzania (photograph by K. Willson).

straight lines as close to the contour as possible. In the past, some very steep land has been planted (Fig. 16.2). Soil-erosion control and access to the plants are difficult on such steep land. In some countries, planting on steep land is forbidden.

PROPAGATION

Seed

Seed orchards (known in the industry as 'baries') are planted with selected trees at a wide spacing (5–6 m apart). They are allowed to grow unchecked, apart from hygienic removal of dead branches and asymmetrical growth, with occasional thinning and removal of the lowest branches; trees must not interlock. The soil below should be kept clean and clear of weeds; seed is collected

Fig. 16.2. Tea on steep land, Sri Lanka (1972) (photograph by K. Willson).

on the ground. Anandappa (1986) discussed the design of polyclonal orchards. Tea-seed bearers require high levels of fertilizer (Anandappa, 1986; TRFK, 1986e).

Tea seed has a short period of viability, so must be planted as soon as possible after collection. If it has to be transported for a long distance, it should be packed in boxes in moist charcoal. The seed must be sorted by flotation immediately before germination. The best sinks within 24 h. Those which sink during the ensuing 72 h may be used but raised separately; there will be a significant proportion of weak plants, which must be discarded. Seeds still floating are discarded.

Before planting in the nursery, seeds must be germinated. They are spread out in the sun, preferably on black plastic, and kept moist by frequent watering; as soon as a seed cracks, it is removed and sown in the nursery. Seeds are usually sown in raised beds of good topsoil, which must have an appropriate pH – soil with a high pH can be treated with sulphur (Table 16.1) – and under

Table 16.1. Treatment of high-pH soil for nursery use (adapted from TRFK, 1986h, by permission of Director, TRFK).

	Sulphur addition (g m^{-3})			
	Minimum		Maximum	
pH	Sulphur	Minimum time between treatment and planting (weeks)	Sulphur	Minimum time between treatment and planting (weeks)
5.0	–	–	–	–
5.1	–	–	60	1
5.2	–	–	115	2
5.3	–	–	170	3
5.4	–	–	225	4
5.5	–	–	285	5
5.6	–	–	340	6
5.7	–	–	395	7
5.8	–	–	450	8
5.9	60	1	510	9
6.0	115	2	565	10
6.1	170	3	620	11
6.2	225	4	675	12
6.3	285	5	735	13
6.4	340	6	790	14
6.5	395	7	845	15
6.6	450	8	900	16
6.7	510	9	960	17
6.8	565	10	1015	18
6.9	620	11	1070	19
7.0	675	12	1125	20

high shade. Planting seeds 12.5 cm apart in a triangular arrangement permits them to grow to 'stumps'. Seeds should be planted with the 'eyes' horizontal and covered by 2.5 cm of soil. Spraying the beds with Simazine pre-emergent weedkiller before sowing minimizes weed growth when the seedlings are small.

Nurseries should be watered regularly so that they do not dry out. Regular applications of foliar fertilizer (usually nitrogen, phosphate and potash) speed the formation of strong plants. Shade should be thinned gradually as the plants grow. Before removal from the nursery, the seedlings should have been in full sun for several weeks to harden off.

Seed may be planted into polythene sleeves if sleeved plants are needed for specific reasons. They may, for example, be better suited for infilling vacancies in fields, where they have to compete with adjacent mature plants.

From nursery beds, seedlings are usually taken as stumps. They are pulled or dug out and cut 10 cm above ground level and 45 cm below. Seedlings are ready to be stumped when they are at least 1 cm in diameter at the collar and have been at least 3 months in full sunlight. A high starch content is vital for survival and growth in the field. Prepared stumps must be kept cool and moist under damp sacking until planted.

Stumps will be ready for planting out after 2–3 years from sowing the seed, depending on growth rate. Sleeved plants can be planted satisfactorily at an earlier age. For infilling, it is better to allow them to grow larger before planting; larger plants should cope better with the competition from established bushes.

Cuttings

Most cuttings root easily in an appropriate medium, although clones vary in their ease of rooting. Cuttings will grow to form good plants if kept moist in the early stages. It has been found that a single leaf cutting is adequate. Larger cuttings, with two or three leaves, will root as easily as single leaves but transpire more in the nursery. The plants formed will produce more branches in the field. Clones are now usually propagated as single-leaf cuttings throughout the industry.

Special mother bushes are kept for production of cuttings. These are planted separately from the producing tea so that they can be given special treatment. Cuttings can be taken from plants in the field but the plants produced may be inferior. Mother bushes should be given fertilizer at a higher rate than the normal fields; for example, double the normal application is recommended in Kenya (TRFK, 1986c). The bushes should be pruned back to stumps at regular intervals, when the stems carry about 12 leaves. At low altitudes, this will be at about every 3 months; at higher altitudes, a longer interval is required, up to about every 8 months. All stems should be cut off; the level of the cut should vary from one cut to the next, around an optimum mean height. Fertilizer applications should be spread evenly throughout the year, with each application halfway between prunes.

Cut stems must be wrapped immediately in damp sacking under shade and watered. Single-leaf cuttings are made by a sloping cut above each leaf. Very long internodes can be shortened by another sloping cut. Finished cuttings must be placed in water until planted in the nursery. The proportion of cuttings that survive falls if they are allowed to dry out (Green, 1962).

Soil has a critical influence on rooting. It must be acid – pH preferably below 5.0 – and have a low content of organic matter. In many areas, the subsoil is very suitable. High-pH soil can be acidified, using sulphur or aluminium sulphate. Table 16.2 lists quantities. In soil that is not sufficiently acid, cuttings will form a callus but no roots.

Cuttings are usually planted into plastic sleeves filled with soil. In most cases, the rooting soil alone is used. Sometimes, the rooting soil lacks nutrients to an extent that restricts the growth of the cuttings. In such cases, a more fertile soil may be put in the lower half of the sleeve, but trials must first establish that roots will cross the interface.

Filled sleeves should be put together in wire frames to make beds about 1.5 m wide. One cutting is placed in each sleeve, with leaf and bud just above soil level. When full, the bed is watered and then covered by a plastic sheet over light frames to keep it clear of the plants. This sheet is sealed into the soil at the edges of the bed so that the plants are totally enclosed. Shade must be constructed at a higher level to reduce the light intensity at the beds. Beds which are not sealed need frequent watering or misting.

The plastic sheets over the beds are lifted a little at a time, starting when the plants have three or four leaves, until the sheets are fully removed. The high shade must be thinned slowly later, so that the plants have been in full sun to harden off before planting in the field. Fertilizer should be applied regularly as a foliar spray as soon as the plants have roots.

Table 16.2. Acidification of soil for field planting (from TRFK, 1986i, by permission of Director, TRFK).

	Sulphur addition	
Soil pH	Broadcast (g m^{-2})	Per hole (g)
5.9–6.4	400	115
6.5–6.9	400	225
7.0–7.4	800	340
7.5 and higher	800	450

Budding and Grafting

Tea can be grafted successfully by several techniques. None is a competitor with cuttings for production for field planting. The technique has been used, firstly, to upgrade seed-bearing trees in orchards by grafting on superior stock and, secondly, in grafting drought-susceptible clones on to drought-tolerant clones for planting in areas regularly suffering from drought.

Propagation *in Vitro*

In vitro propagation of tea (tissue culture) has not been found to be easy. There are, however, reports of some success in some laboratories (e.g. Manivel, 1993; Banerjee and Agarwal, 1990; Sarathchandra *et al.*, 1990; Tavartkiladze, 1990). Both direct and somatic embryogenesis has been achieved. It is very unlikely that this method will have any value for large-scale commercial planting, but it has already become a tool for research. Enhancement of rooting of *in vitro* shoots was reported by Jain *et al.* (1993). The subject has been reviewed by Tahardi (1994).

FIELD PLANTING

Spacing

For high production, it is important that a tea plantation covers the ground completely. The soil should be almost totally shaded so that its temperature is kept low; this helps to maintain the mulch layer, which is vital for efficient nutrition, and weed growth is minimized. Field-planting techniques must therefore have the aim of developing a full ground cover as quickly as possible. A wide range of spacings has been used over the years. In recent years, closer spacings have become more common, in order to achieve full ground cover more quickly. Magambo and Waithaka (1985) studied the influence of plant density on the yield of young clonal tea.

Other factors can have a bearing on spacing. Clones have a range of growth habits and it may be desirable to plant clones with a narrow natural spread closer than clones with a wide-spreading habit. Machine harvesting can dictate the optimum interline distance.

A very wide range of spacings, or plant densities, has been tried on an experimental basis, from around 5000 to over 400,000 plants ha^{-1}. In general, a parabolic relationship between spacing and yield per unit area of land is found (Rahman *et al.*, 1981) (Fig. 16.3). The spacing giving the highest yield reduces as the tea plants age; Rahman's data show that it falls from over 68,000 plants ha^{-1} 2 years from planting to 17,000 5 years from planting.

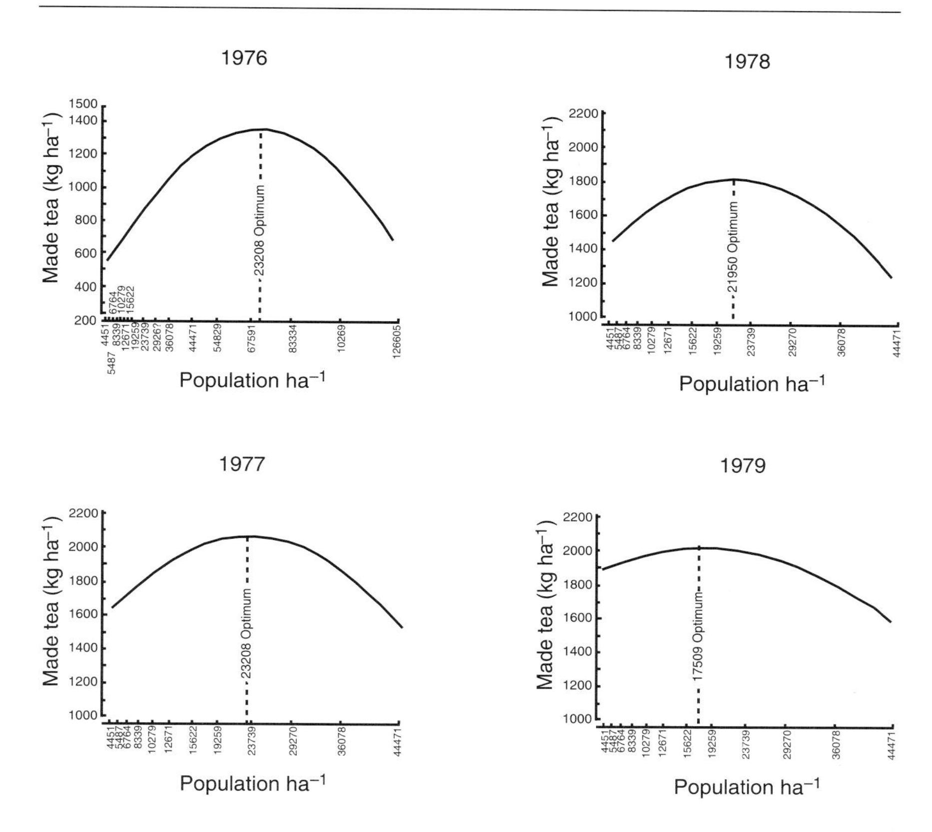

Fig. 16.3. Relationship between yield of tea and planting density for the first 4 cropping years (from Rahman *et al.*, 1981).

Yield from mature tea is at its highest at spacings between 10,000 and 15,000 plants ha^{-1}. It is probable that most of the tea planted in recent years has been within that range – for example, hedge plantings, typical spacings being 1.2 m between hedge lines and either 0.6 m (requiring 13,889 plants ha^{-1}) or 0.75 m (11,111 plants ha^{-1}) between the plants in the hedge.

Planting

The site for each plant should be marked by a stick without disturbing the cover crop. Planting holes should be dug out beforehand and the soil left alongside. It is desirable in most cases to add fertilizer to the soil, depending on the soil analysis – usually a phosphate and sometimes potash or a suitable and locally available organic manure (TRFK, 1986d). The fertilizer should be well mixed with the soil from each hole before planting the tea. If the soil has a high pH, sulphur should be mixed in also, to lower the acidity, if this has not been applied broadcast. Quantities are listed in Table 16.1.

Stumps require a hole 60 cm deep and 20–30 cm in diameter; they should be planted so that the field soil is at the same level as the nursery soil. Sleeved plants need a smaller hole, about 45 cm deep and 25–30 cm diameter. Allowance must be made when backfilling for subsidence in the hole. If subsidence leads to a saucer, the plant may become damaged if water stands around it.

After planting, the marking stick should be put in at the side of the plant at an angle so that it goes over the plant to indicate its position. In the absence of a cover crop and even with one, it may be necessary to shade young plants with leaves or other suitable material of a temporary nature. In some areas, a grass mulch is recommended; this should be spread as soon as possible after planting.

Infilling and Interplanting

Bushes inevitably die for a variety of reasons in established tea fields. Weeds grow in the gaps and a significant number of vacancies reduces overall production. It therefore becomes necessary to replace missing plants, an operation known as 'infilling' or 'supplying'.

Larger plants than those used for normal planting are desirable to help them compete with neighbours. A large hole should be dug so that there are no live roots close to the new plant. Fertilizer must be mixed with the soil in the hole. Where several adjacent plants in a row have died, new plants are often put in at a closer spacing than the original plants in order to speed the return of total ground cover. The same technique is used if it is desired to put additional plants in tea planted initially with a very wide spacing.

SHADE

When the tea industry started in Assam, the tea plants were always planted under the shade of various trees. It was later found in many areas that tea under shade had a limited response to fertilizer applications, while unshaded tea had a better response to fertilizers and could produce much higher yields; shade trees were therefore removed in many areas. Othieno and Ng'etich (1993) reported that, at a high altitude in Kenya, unshaded tea bushes produced significantly more crop-bearing shoots than shaded plants. However, unfertilized shaded tea yields better than unfertilized unshaded tea. Therefore, in low-input systems, shade trees could be an advantage. In some areas, temperatures are very high in the peak growing season and yield falls because tea-leaf temperatures exceed the critical value and photosynthesis slows down (see Chapter 15, p. 180). In these areas, such as the low-altitude areas of north

India and Bangladesh, shade trees are retained. Hadfield (1974a, b) studied shade and the distribution of light under various shade patterns.

Where shade trees are required, they should be planted before the tea, preferably during the previous year, so that they are well established before the soil is disturbed by planting tea. Shade trees can be raised from seed on the plantation, or it may be possible to purchase seedlings from a forest nursery. Most of the species that have been used for shade are leguminous, particularly a number of *Albizzia* species. Leguminous trees can provide a significant amount of nitrogen, fixed from the air, for the tea. *Grevillea robusta*, which is not a legume, has also been widely planted as a shade tree. Shade trees have to be managed by careful pruning, so that an even, dappled shade is formed on the tea below.

Shaded tea is shown in Fig. 16.4 and unshaded tea in Fig. 16.5.

SHELTER

Where tea is exposed to strong winds, shelter-belts of lines of suitable trees can be planted, but there are disadvantages to this practice. Lines of shelter trees should be at a right angle to the direction of the most troublesome wind and separated from each other by a distance ten times the height of the trees above the tea. They will reduce the yield of tea in adjacent lines by competing with and shading the tea plants. Such losses must be weighed against those caused by the unchecked winds (TRFK, 1986a).

Fig. 16.4. Tea under shade, Assam, India (photograph from Tea Research Association, India).

Fig. 16.5. Unshaded tea, Australia (photograph from Madura Tea Estate, by permission of the owner).

Hakea saligna has often been planted as a shelter tree; its height is limited and it has everlasting foliage over a convenient range of height above the tea. Many of the trees used for shade have also been used for shelter. Taller trees, such as *Eucalyptus* spp., which are unsuitable for planting within tea fields, may be planted in lines some distance from the fields to reduce the strength of the wind.

BRINGING INTO BEARING

It is desirable, having planted tea, to bring it to a condition where crop can be harvested, as quickly as possible. This process is known as 'bringing into bearing'. The plants must be made to spread so that they touch and form a complete ground cover of crop-bearing stems. The lower branches formed during this process form the 'frame' of the plant and carry the maintenance foliage, which is vital for photosynthetic production of carbohydrates (Manivel, 1978).

Bringing into Bearing by Pruning

When a bush is cut back, new stems will grow from leaf axils below the cut, some of which will grow outwards from the bush and increase its spread. Young sleeved plants from the nursery usually have only one stem, although removal of the apical bud after about five leaves have formed can stimulate some branching. Stumps usually form three to six branches after planting in the field. In either case, it is often necessary to cut across plants in the field twice; the second cut will be at a higher level than the first, so as to retain the spread induced by the first cut. Typically, bushes will be cut at 20 cm and 40 cm above ground level. There will therefore be a delay of about 3 years between planting and the first harvest.

Bringing into Bearing by Pegging

An alternative to pruning is to peg down the first stems formed so that they are almost horizontal and evenly spaced around the centre. New vertical stems will arise from buds on the upper side of the pegged stems, so long as no part of the stem dips downwards. The new stems grow upwards and will carry the first crop when they reach the appropriate height. This method avoids the delays caused by pruning and regrowth. A substantial saving in time, from about 3 to around 2 years, is usually possible. Young plants should preferably have at least four stems before pegging. It is better to ensure that these are formed in the nursery by removing the apical bud. If a single-stem plant is pegged down, new stems will arise along its length, but these will not be symmetrically spaced around the centre of the plant.

Figures 16.6 and 16.7 outline the two methods. A discussion of experimental results on pruning and pegging is to be found in Tea Research Institute of East Africa (1968). Figure 16.8 shows a young plant with stems pegged down.

Tipping

This process is the removal of the bud and the first two or three leaves of a shoot in order to level the tops of all bushes at the required height.

It is possible to bring plants into bearing by tipping at several levels instead of pruning – for example, 20 cm, 30 cm and 40 cm above ground level on successive occasions. The new branches formed below tipping do not usually

Fig. 16.6. Bringing into bearing by pruning. 1. Young plant pruned at a height of 20 cm. 2. The same plant after the second prune at a height of 40 cm. 3. The same plant after tipping to form a plucking table at 60 cm. (From Tea Research Foundation of Kenya, by permission of the Director.)

1

2

Fig. 16.7. Bringing into bearing by pegging. 1. Young plant, with four stems pegged down. 2. Mature plant which had been pegged down as in 1 above. (From Tea Research Foundation of Kenya, by permission of the Director).

spread as well as those from a lower pruning cut. More frequent removal of the upper stems may weaken the bushes. Nevertheless, the bushes may be up to normal plucking height more quickly than with other methods and some crop is given on the way, as leaf removed during tipping can be processed. Satyanarayana (1991) investigated several tipping routines and found two two-stage systems which were more productive than the single-stage routine in current use.

Bushes which have been pruned or pegged are tipped when the new stems are a little higher than the initial harvesting level. The objective is to form the level 'plucking table', which is the basis of harvesting. At tipping, a level table is more important than the material removed, which will have various numbers of leaves with the bud. Such material can be processed, provided it is only a very small proportion of factory input.

Fig. 16.8. Young tea plant which has been pegged down to spread the branches, from which vertical shoots will form the frame of the plant. Kenya. (Photograph by K. Willson.)

HARVESTING

The young shoots of the tea plant contain the highest concentration of the substances necessary to make the tea flavour. The concentration is highest in the bud about to open and falls as the leaves age; it is even lower in the stem. In order to obtain the highest production and a good-quality product, it is necessary to compromise over the number of leaves harvested and the length of stem included. It has been established that two leaves and the bud is best but three leaves and a bud is acceptable for some purposes. Modern factories can separate the higher-quality fraction of the made tea.

The aim of harvesting is therefore to remove all the shoots which have produced either 'two and a bud' or 'three and a bud', while leaving immature shoots to develop. It may be necessary to harvest some shoots that are 'banjhi', i.e. dormant, removing the bud and one or more leaves to stimulate growth of a new shoot. The crop must be harvested at intervals throughout the growing season. The length of the interval between harvests varies according to the growth rate (see Chapter 15). The crop must be harvested to give the maximum amount of 'two and a bud', or 'three and a bud' with a minimum of overgrown shoots.

Harvesting is known as 'plucking', as selected stems of the correct size are plucked off the bushes (Fig. 16.9). Each time a field is plucked, it is called a

Fig. 16.9. Hand plucking a tea shoot, East Africa (photograph by K. Willson).

'plucking round', and the interval between plucking rounds is known as the 'plucking cycle'. It is convenient for plucking to have a smooth surface at the top of the bushes. In many countries the top is flat, approximately parallel to the ground. This is known as the 'plucking table'. In some other countries, notably Georgia and Japan, the top of the bushes in a line is cut to a circular shape. The smooth surface, flat or curved, enables the pluckable shoots to be seen.

The severity of plucking has a significant effect on crop yield. As the severity is increased, the harvested shoots are removed from closer to the mother leaf from which the shoot has grown. Thus 'hard' plucking removes shoots from close to the mother leaf and the plucking becomes softer as the point of plucking rises, firstly to above the scale leaf and then above the fish leaf.

A balance has to be set between ensuring a high proportion of the desirable two (or three) and a bud and maintaining an adequate and efficient maintenance foliage. The leaves in the maintenance foliage must be renewed regularly; this is done by allowing the plucking table to rise periodically. Young leaves are added to the top of the maintenance foliage, while the oldest leaves, at the bottom, become senescent and fall off. Sharma and Satyanarayana (1993) studied the interactions between the level of the plucking table above the pruning height, the plucking cycle, the severity of plucking and the quality of the plucked leaf. They showed that it is possible to regulate these factors so that non-selective plucking, mechanically for example, can produce high-quality leaf at certain times of year.

Fig. 16.10. Manual harvesting ('plucking') of tea, Nilgiris district, India (photograph from the Tea Council, London).

Manual Harvesting

The harvesters, 'pluckers', move along between lines of tea, removing all the shoots which are ready and putting them into a basket or other container, which is often carried on the back (Fig. 16.10). A long, straight stick is often laid on the plants as a guide to the level. In recent years, in some plantations, the actual plucking has been replaced by shears. These incorporate some form of collecting device, which holds the cut leaf until it can be emptied into the basket. These are inevitably non-selective, so it is important to manage the tea to give a very high proportion of the required quality at the time of plucking.

Programmes to improve the quality of plucked leaf have recently been introduced. 'Scheme plucking' allocates one plucker to a particular area of tea. This person plucks this tea on every occasion it is necessary. This is extended by 'programmed plucking'. In this scheme, the dates on which each field is to be plucked are calculated in relation to the rate of growth expected from the normal weather at the time of year.

Murthy (1992) surveyed manual harvesting practices in South India by hand or manual shears. The effects on crop yields are discussed and costs estimated.

Mechanical Harvesting

Plucking is labour-intensive, so interest in mechanization has grown over the years. Many trials have been carried out to compare costs and product quality from hand and mechanical plucking. With careful control in the field and a modern factory, there need be no great loss of quality production. Barbora *et al.* (1993) showed that a wheel-mounted harvester was the most cost-effective plucking system at times of peak yield. Huang and Chiu (1990) discussed the replacement of manual plucking by machines in Taiwan.

Several types of machine have been developed. The first is a mechanized shears, which can be operated by one person. Next in size is a reciprocating cutter, long enough to cover one row of tea. This is mounted on a frame so that it can be carried by two persons, who walk one each side of the row of tea. This type mounted on wheels and pushed by the operators becomes a 'handcart'. All the foregoing have to include some arrangement for collecting the cut leaf.

Finally, there are self-propelled harvesters of varying degrees of sophistication. These straddle one or two rows of tea. Cutters can be reciprocating or rotary, while trials have been made of simulating plucking, using rubber fingers. Such machines are mounted variously on three, four, six or eight wheels or are half-tracked or fully tracked. Cut leaf must be removed expeditiously from the cutters or it will be damaged. Transport within the machine is usually either by a current of air or by a moving belt. Gabuniya (1991) described and assessed a number of machines developed in the former USSR. Figures 16.11 and 16.12 each show a mechanical harvester which is in regular use. Kilgour and Burley (1991) described the machine developed at Silsoe College for Uganda. While the area of tea harvested mechanically is at present relatively small, it is increasing. Nixon and Burgess (1996) studied the effects of mechanical harvesting on the yield and leaf quality of clonal tea. Melican (1997) discussed the relative values of hand and mechanical plucking.

Fig. 16.11. Self-propelling half-track mechanical tea harvester developed by Silsoe College, Cranfield University, England (Kilgour and Burley, 1991). The cutting head is to the right, in the box below the main frame. Cut tea is carried thence to the basket at the rear. (Photograph from Silsoe College.)

Handling of Cut Leaf

Cut leaf starts to dry ('wither') and also to ferment as soon as it is cut. Leaf quality deteriorates as these processes continue in an uncontrolled manner. Stacking leaf too deeply causes crushing, which also causes deterioration. It is important to get leaf to the factory as soon as possible, so that the essential withering and fermentation can be controlled.

Plucked leaf is weighed at the field and then moved to the factory. A variety of means of transport have been used over the years. Vehicles should be designed to avoid crushing. Intermediate flooring is essential to separate different layers in the load.

PRUNING

Maintenance Pruning

The upper part of the plants, below the plucking table, becomes congested, with stems and knots where many new stems have originated. Also, the lower foliage, the 'maintenance foliage', ages and loses efficiency, lower leaves dropping off. Crop

Fig. 16.12. Mechanical tea harvester developed by Madura Tea Estate, Australia. The cutting head for harvesting is on the left. The inclined tubes carry the cut tea pneumatically to the basket on the right. Between the wheels below the frame are rotary cutters, which can be fitted when required. (Photograph from Madura Tea Estate.)

yield can therefore diminish. This effect is countered by allowing the plucking table to rise slowly, adding new leaf to the top of the maintenance foliage. Ultimately, the plucking table becomes inconveniently high and the maintenance-foliage efficiency very low. At this stage, the bushes must be pruned.

The bushes are cut across at a fixed level, above the basic structure of the bush frame, but into mature branches. Some sanitary pruning may be done at the same time, removing unwanted, diseased and weak stems. This pruning is done at intervals which vary from 1 to 5 years, depending on growth rate and climatic conditions. This is the 'pruning cycle'. The pruning level is usually a little higher than the previous prune each year, to avoid excessive knotting. Figure 16.13 shows mature tea plants after maintenance pruning; at the level shown, it is a cut-across prune. Cut across at a lower level, much less maintenance foliage remains on the stems; this is called a 'clear' prune.

New stems grow from the stems remaining. These are tipped at the appropriate level to form a new plucking table and the cycle is repeated.

Some stems may be left uncut when the remainder are cut off. These are known as 'lungs'; they give some shade to the cut stems and continue photosynthesis, which helps to speed recovery. The lungs are removed when new stems are sufficiently advanced.

Fig. 16.13. One line of tea bushes which has been given a high-level maintenance prune. Note the flat top surface and the extra height of adjoining rows. This level of prune is almost as high as a skiff. Location not recorded. (Photograph by K. Willson.)

'Down', 'Collar' or 'Rejuvenation' Pruning

After many years, the basic frame ages and it is not possible to carry out a maintenance prune at a sufficiently low level. The bushes then have to be cut very much lower, often into the main stem below the frame, the 'collar'. New stems grow from the collar and the bringing-into-bearing process has to be repeated. There are usually vacancies where bushes have died by the time that this process is necessary. Njuguna and Magambo (1993) showed that rejuvenation pruning and interplanting with clonal plants did not result in as high a yield as complete replanting with clones.

'Skiffing'

This is a light prune at a relatively high level into mainly immature branches. It is done sometimes in order to lower the plucking table to extend the pruning cycle. Recovery is quicker than from a normal maintenance prune, so there is a smaller loss of crop. For detailed discussions of pruning techniques, see Kulasegaram (1986) and TRFK (1986b). Hudson (1992) discussed pruning, its effect on crop yields and its costs in South India. Hazarika *et al.* (1965) studied the effects of extending pruning cycles in Darjeeling.

Mineral Nutrition and Pruning

Recovery from pruning will be slow and uneven if the bushes are short of nutrients. In addition to the normal fertilizer applications, extra quantities are often given in the year of pruning. This is often a 'remedial' application to counter mineral deficiencies which have been diagnosed.

Prunings and Nutrition

The branches cut off at pruning contain very large quantities of nutrients. They must not be removed from the field. They should be left to rot down and form a mulch layer, in which the nutrients will slowly be released to be reused by the plants. The mulch layer is very important as the source and site of nutrient absorption; it is filled with a dense mat of roots, which can be seen by lifting the upper layer of mulch (see Fig. 15.5, p. 189). It also restricts weed growth. Some prunings should be put over the plant frames after pruning as a partial shield against sun scorch.

17

Mineral Nutrition and Fertilizers

NUTRIENTS REMOVED IN CROP

A number of writers have reported quantities of nutrients removed from the field in the tea crop. These figures are based on a crop yield and nutrient content of the leaves in the crop. For a crop yield of 2500 kg made tea ha^{-1} and using the minimum nutrient contents of the third leaf (dried) on the plant, the amounts removed in the crop are: nitrogen (N) 100 kg ha^{-1}, potassium (K) 50 kg ha^{-1}, which equals 60.25 kg K_2O ha^{-1}, and phosphorus (P) 10 kg ha^{-1} equivalent to 22.9 kg P_2O_5 ha^{-1}. Therefore substantial quantities of nitrogen and potash are needed, merely to maintain a crop level which is common and often exceeded.

CIRCULATION OF NUTRIENTS

It is important to understand the way nutrients are recycled in a tea plantation. This is shown schematically in Fig. 17.1. Of particular importance are, firstly, the large quantity of nutrients that are contained in the prunings and leaf fall – almost four times the quantity removed in the crop – and secondly, the small amount of nutrients absorbed from the subsoil and the large quantity absorbed in the mulch that has been formed from prunings and leaf fall. These figures underline the importance of retaining all prunings in the field. If prunings are removed, either the tea will soon become severely deficient in, particularly, nitrogen and potassium or much larger applications of fertilizer will have to be made. Thirdly, there is a need to apply potash fertilizer in addition to the NPK 25:5:5 which has been in common use for many years.

Fig. 17.1. (Opposite) Flow diagram of nutrients in a tea plantation over a 3-year pruning cycle.

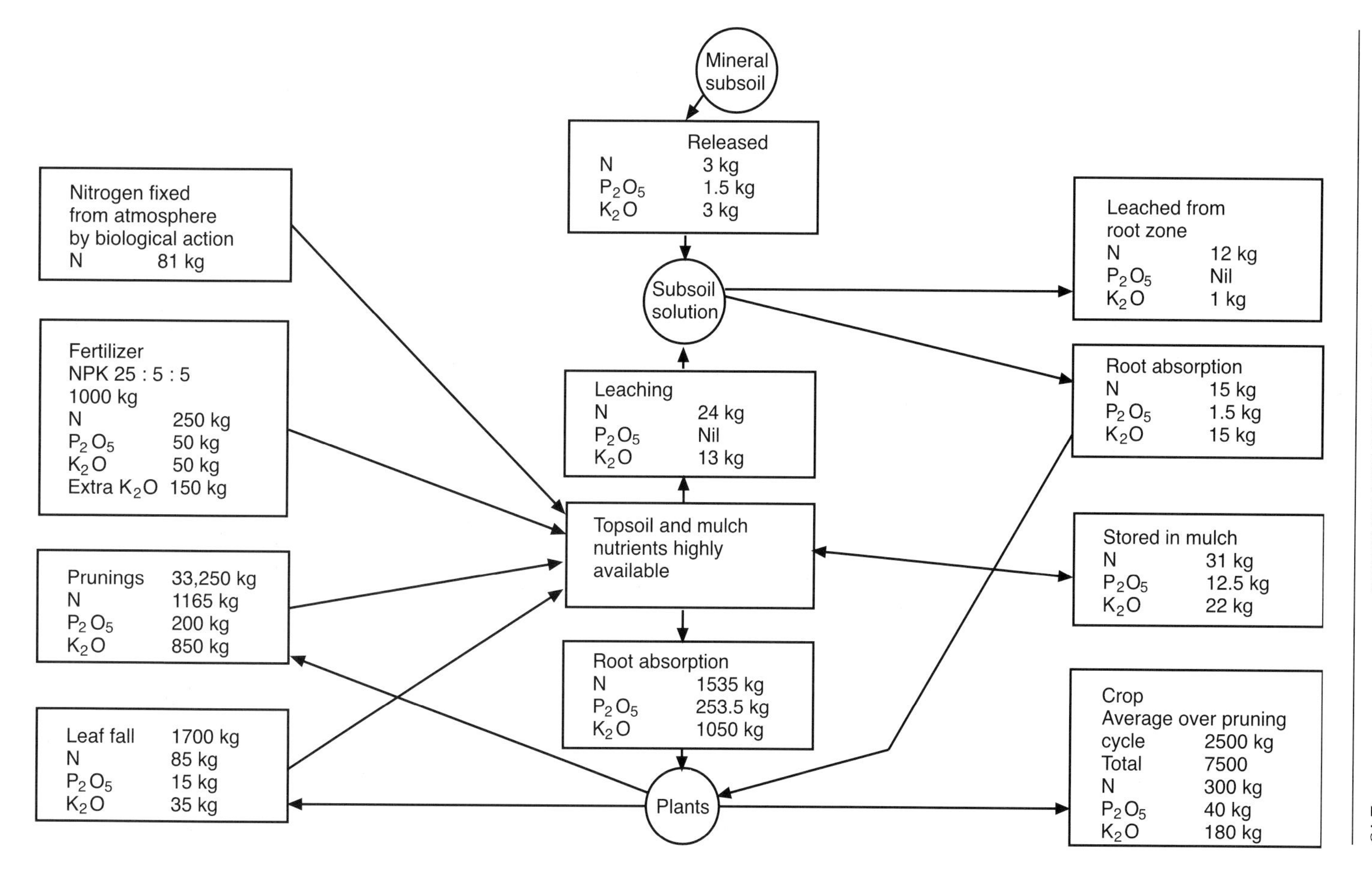
Mineral subsoil
Released
N 3 kg
P_2O_5 1.5 kg
K_2O 3 kg
Nitrogen fixed from atmosphere by biological action
N 81 kg
Subsoil solution
Leached from root zone
N 12 kg
P_2O_5 Nil
K_2O 1 kg
Fertilizer
NPK 25 : 5 : 5
1000 kg
N 250 kg
P_2O_5 50 kg
K_2O 50 kg
Extra K_2O 150 kg
Leaching
N 24 kg
P_2O_5 Nil
K_2O 13 kg
Root absorption
N 15 kg
P_2O_5 1.5 kg
K_2O 15 kg
Topsoil and mulch nutrients highly available
Prunings 33,250 kg
N 1165 kg
P_2O_5 200 kg
K_2O 850 kg
Stored in mulch
N 31 kg
P_2O_5 12.5 kg
K_2O 22 kg
Root absorption
N 1535 kg
P_2O_5 253.5 kg
K_2O 1050 kg
Leaf fall 1700 kg
N 85 kg
P_2O_5 15 kg
K_2O 35 kg
Plants
Crop
Average over pruning cycle 2500 kg
Total 7500
N 300 kg
P_2O_5 40 kg
K_2O 180 kg

Swaminathan (1992) discussed integrated nutrient management, which has to keep in mind the factors mentioned above. Verma and Sharma (1995) reported cost-effective nutrition for tea in South India.

EFFECTS OF INDIVIDUAL NUTRIENTS

Major Nutrients

Nitrogen

Nitrogen is vital for a crop that consists of leaves. It accounts for up to 5% of the dry weight of the crop. Application of nitrogen fertilizer has been part of tea husbandry for many years.

Experiments over a long period have shown that application of nitrogen gives a response in yield which is linear to a high level. The response depends on there being an adequate supply of all other nutrients. The response is usually of the order of 4–8 kg of made tea for each kilogram of nitrogen applied. The response usually begins a few weeks after application if the tea is growing well. It has also been found that the response to a given application increases year by year for at least 10 years. The response is likely to be highest in the second year of a pruning cycle, declining thereafter. Very heavy pruning, particularly collar pruning, often stimulates the nitrogen response for several subsequent pruning cycles. Nitrogen applications in high-yielding tea are often over 200 kg ha^{-1} per annum. Such high levels are usually split over several applications, but Nyirenda (1994) showed that there was no advantage in applying nitrogen in three portions to rain-fed tea during the Malawi rainy season, compared with one application. Malenga (1996) reported the responses of young clonal tea to high levels of nitrogen. The response to nitrogen is restricted in many crops if the crop lacks potassium. The potassium–nitrogen interaction is particularly important in tea (Gething, 1993). Bezbaruah *et al.* (1993) showed that ureolytic nitrifying bacteria stimulate the uptake of nitrogen by tea plants. Sandaman *et al.* (1978) studied the nitrification of nitrogenous fertilizers in acid yellow-podzolic soil.

The effects of nitrogen fertilization levels on the amounts of the chemicals responsible for the taste of the tea were investigated by Whitehead and Muyila (1992).

Phosphorus

This element is indispensable for the formation of new stems and roots and in several processes within the plant. Tea does not need a large quantity but often has difficulty in absorbing the optimum amount. Most phosphorus is fixed in typical acid tea soils. Direct responses to applications of fertilizer to mature tea are rare; interactive responses with another nutrient are probably more common. Despite the lack of direct response, it is good practice to apply a small quantity of phosphate fertilizer regularly. A slow increase in the amount of phosphorus in the leaves has been reported and direct responses have been noted after several years of regular application. The mulch which should build up under mature tea should help this benefit; there is less fixation of phosphate in organic media. It has been shown that vesicular-arbuscular (VA) mycorrhiza play a part in phosphorus absorption by tea roots (Neog, 1986; Zhi, 1993).

In contrast, phosphate can have a major effect on young plants in nurseries and on planting in the field. It is useful to incorporate some phosphate in nursery soil, and foliar application to nursery plants hastens growth. A mixed NPK fertilizer is usually used. Large benefits have followed the incorporation of phosphate in the soil from planting holes before field planting (O'Shea, 1964). The phosphate must be well mixed with the soil; direct contact between particles of fertilizer and young plants can produce severe adverse effects.

Applications of phosphate to mature tea often exceed 100 kg ha^{-1}.

Potassium

Potassium is a vital nutrient, but its role in the plant is not fully understood. Tea suffers severely when it cannot absorb sufficient potassium. Application of potash fertilizer to tea showing symptoms of potassium deficiency produces a dramatic improvement very quickly. As acid soils do not contain large quantities of available potassium, application of potash is essential to almost every field of high-yielding tea. Currently, applications often exceed 200 kg K_2O ha^{-1} per annum. Uptake is significantly increased by the VA mycorrhizal fungus *Glomus fasciculatum* (Zhi, 1993).

Potassium has a marked effect on recovery from pruning. Frequently, additional potash fertilizer is applied not long before or after pruning. This is in addition to regular applications as part of a routine NPK fertilizer programme. The need for this additional fertilizer is often indicated by low levels of potassium in the leaves when analysed before pruning. Where analysis shows a deficiency, a 'remedial' application of potash fertilizer is made.

Potassium and calcium are antagonistic in tea. When there is an excessive amount of calcium in the soil, this is absorbed in above-normal quantity and blocks the uptake of potassium. Severe potassium deficiency results. Application of potassium fertilizer depresses calcium uptake (Shuvalov and Sardzhveladze, 1990).

Magnesium

This element is essential, particularly as a constituent of chlorophyll, but the proportion in tea leaves is relatively small. Magnesium is very mobile within the plant, so that it is moved to younger leaves from older ones when supplies are deficient. Deficiency symptoms therefore appear initially on the oldest leaves and the effect on crop yield is marginal, as the oldest leaves do not carry out much photosynthesis, due to lack of light. Dolomitic limestone is sometimes applied to soils with low pH and low magnesium content.

Calcium

A small quantity of this element is essential. A deficiency of it is very rare, as even highly acid soils contain a little calcium. Large amounts of calcium inhibit potassium uptake and a deficiency results.

The application of lime to acid tea soils usually depresses crop yield. In some cases where very acid soil contains substantial amounts of potassium, lime can be applied occasionally and improves the pH, with no adverse effect on crop yield. Applying large quantities of fertilizers containing calcium, such as some phosphates and calcium ammonium nitrate (CAN), often depresses yield.

Sulphur

This element is essential as it takes part in important chemical reactions in the leaves and forms part of several important groups of compounds. Sumbak (1983) discussed the sulphur requirements of tea.

Some tea soils lack sulphur and deficiency symptoms appear on young plants. Planting-hole application of a sulphur compound prevents the deficiency occurring in young plants in the field. Sulphur is a constituent of many fertilizers, particularly sulphate of ammonia (SA) and single superphosphate (SSP). Sulphur deficiency is therefore very rare in mature plants. When fertilizers not containing sulphur, such as urea, are used, care must be taken to include a sulphur-containing material in the programme.

Minor Nutrients

Iron

Iron in small quantity is vital. It is usually freely available in acid soils, so deficiencies are almost unknown. *Glomus fasciculatum* on the roots improves the uptake of iron (Zhi, 1993).

Manganese

With iron, this element seems to be involved in chlorophyll formation without being included therein. It is usually freely available in acid soils, from which tea absorbs large quantities. The amount in leaves must reach a very high level before any toxic effect on growth is seen.

Boron

Also vital in small quantity for several processes in the plant, this element concentrates at growing points and in flowers. Most tea soils can provide adequate boron, so deficiency problems are rare. Boron uptake is reduced as the amount of calcium in the soil is increased (Malita and Mahanta, 1993).

Both positive and negative effects on uptake of calcium have been reported. In the absence of boron, root growth but not shoot growth is stimulated. Pollen tube growth was inhibited; addition of fluorine overcame this effect. Other species grown alongside tea, such as some shade trees, show deficiency symptoms when tea is not affected (Smith, 1960).

Copper

In addition to taking part in the synthesis of chlorophyll and vitamins this element is a constituent of polyphenol oxidase, which catalyses the fermentation process in tea. When copper is deficient, the tea will not ferment properly and the quality of the product is reduced. A mild deficiency is quite common but can be rectified by foliar application of a copper compound. *Glomus fasciculatum* in the roots improves the uptake of copper (Zhi, 1993).

Zinc

This vital element takes part in many essential processes, particularly the hormones and auxins that control plant growth. Deficiencies of zinc are common in tea with high yield potential. Treatment must be by foliar spray of a zinc compound, as uptake of soil-applied compounds is very poor. Boron in excess inhibits the uptake of zinc.

Aluminium

A small quantity of aluminium speeds the assimilation of phosphorus and potassium, thus improving plant growth. It is freely available in many acid soils and tea accumulates large quantities, which have no direct effect on growth. Older leaves contain up to three times the amount of aluminium contained in the young flush (Hasselo, 1965). Reduction of uptake of calcium and magnesium has been reported with no significant effect on crop yield. Tsuji *et al.* (1994) studied the stimulant effects of aluminium sulphate and phosphate fertilizer on cultured tea roots. Applied together, the two elements stimulated growth in acid soils down to a pH level of 4.0; applied separately, growth was not stimulated. Growth stopped in the absence of aluminium.

Konishi (1990) reviewed the effects of aluminium on plant growth. Owuor *et al.* (1990) reported aluminium levels in clonal teas.

NUTRIENT IMBALANCE SYMPTOMS

Nitrogen Deficiency

The colour of young leaves lightens and can become almost white; the colour loss is most marked in direct sunlight. The growth of stems is reduced, giving shorter internodes. Older leaves lighten and then brown and fall.

Phosphorus Deficiency

Leaves undergo a subtle change, losing the normal, typical, gloss and becoming dull and matt. Woody stems die back, which can be confused with sunscorch, but in this deficiency dieback starts at ends which have been cut when pruning.

Potassium Deficiency

Leaves develop marginal scorch and may turn reddish bronze. Old leaves discolour first and become loosely attached to the stems; shaking the bush will make many leaves fall. Defoliation of the maintenance foliage leaves only a thin layer of leaves at the top of the bush. Growth and bud break slow up and many banjhi shoots with small leaves are formed from the upper sections of stems. Bark becomes silvery-white, while brown blight (*Colletotrichum coffeanum*) and grey blight (*Pestalotia theae*) appear on leaves. Lack of bud break lower in the bushes results in plants not spreading after pruning. Application of potash fertilizer promotes rapid bud break and restoration of the bush framework.

Sulphur Deficiency

Leaves turn bright yellow between the veins, which remain green. On new growth, internodes shorten and leaves become smaller. Leaves scorch and fall off and stems die back. This is known as 'tea yellows'.

Calcium Toxicity

Where calcium is present in great excess, as on soils with a high pH, growth slows down. Internodes shorten, while leaves do not reach full size before yellowing and curling backwards. Leaf edges blacken and leaves distort, crack and fall off.

Zinc Deficiency

Two or more small buds form in place of one to form a 'rosette'. Leaves elongate and twist to a sickle shape and may develop a wavy edge and yellow margins. On new stems, the internodal distance becomes very small.

Copper Deficiency

A slight darkening of leaves in the field is difficult to detect. The main effect of copper deficiency is seen in the factory, where fermentation is inhibited, leaf becoming dark brown rather than bright orange.

Manganese Deficiency

This rare deficiency occasionally occurs on high-pH soils. Leaves become pale to yellow on the edges, with a mottling of red-brown spots on the lamina.

Manganese Toxicity

A toxicity can develop with very high levels of manganese in leaves, which harden, turn dull brown and become very brittle.

Boron Deficiency

Rarely seen in tea, the apical buds fail to grow and and cork develops on the upper side of the leaf petiole.

CONTAMINATING ELEMENTS

Caesium

Unlu *et al.* (1995) investigated the amount of caesium-137 (^{137}Cs) in tea contaminated by the Chernobyl disaster. They reported a rate of decay which reduced the amount of radioactivity by 94.8% in 125 days.

Fluorine

Sha and Zheng (1994) reported the fluorine content of fresh leaves of tea planted in Fujian Province of China.

Selenium

Shuvalov and Konidze (1987) studied the uptake of this element by tea.

Other Elements

A list of amounts of 12 other elements in dry black tea is given in Willson (1991).

MINERAL FERTILIZERS IN REGULAR USE

Nitrogen

Sulphate of ammonia has been used for many years in tea. It provides sulphur, in addition to nitrogen. It acidifies the soil, a property which is very useful in many situations. After many years of use, soil can become very acid, so a change of fertilizer is advisable. In very acid conditions, deficiencies of bases are probable; application of base fertilizers will help to offset high acidity.

Ammonium nitrate (AN) has a higher nitrogen content than SA but is too difficult to handle to be used on its own in wet tropical conditions. It contains no sulphur. It is used at times mixed with SA as ammonium sulphate nitrate (ASN). Calcium ammonium nitrate (CAN), in which it is mixed with lime, has been tried on tea, but the lime content causes problems with excess calcium. AN is also used as a constituent of compound fertilizers, with phosphate or potash compounds or both.

Urea has a high nitrogen content but no sulphur. Nitrogen can be lost from urea if it is applied in inappropriate climatic conditions – high temperature without rain or with excessively heavy rainfall. It is also used as a constituent of compound fertilizers.

Phosphate

Single, double and triple superphosphates (SSP, DSP, TSP) are used on tea. Single has the lowest P_2O_5 content, but does include sulphur. Most other phosphate materials have high calcium contents, which could be troublesome in the long run. The superphosphates are incompletely soluble in water, so are not suitable for foliar application. Monoammonium phosphate (MAP) and diammonium phosphate (DAP) are both soluble in water and also contain nitrogen. They are useful for foliar application and may be used in NPK materials.

Potassium

Muriate of potash (MoP) (potassium chloride) is the potassium material in most common use, both on its own and as a constituent of NPK materials. It contains no sulphur.

Sulphate of potash (SoP) contains sulphur but is more expensive than MoP. It is used on its own and as a constituent of NPK materials, where its sulphur content is necessary.

Magnesium

Epsom salt (magnesium sulphate) is soluble in water and therefore suitable for foliar application. It is used both on its own and in compound materials, where magnesium is needed in addition to NPK; it can also be a constituent of spray formulations of minor nutrients.

Kieserite is an impure form of magnesium sulphate. Insoluble impurities make it unsuitable for foliar application.

Dolomite (magnesium limestone) is an insoluble material which also contains a lot of lime. It is sometimes used on acid soils and releases calcium and magnesium slowly. The high calcium content could be troublesome in tea.

Zinc

Zinc sulphate is soluble in water and can be obtained as monohydrate or heptahydrate. The former has the higher zinc content. Zinc oxide is insoluble but can be used for a foliar spray as a finely ground powder in suspension.

Copper

Copper sulphate is soluble in water but can scorch leaves in certain conditions. Copper oxychloride is insoluble, but finely ground material can be sprayed in suspension. This material is intended primarily as a fungicide but is equally effective as a foliar source of copper.

ORGANIC FERTILIZERS

Many organic fertilizers are, or have been, used on tea. In general, they have relatively low, and often variable, contents of nutrients. Their use therefore involves a lot of handling, so that they are only economic if they are available at minimum cost close to the plantation.

Green manure can be cut from vegetation, either natural or planted for the purpose, and spread as a mulch around the tea plants. This operation is labour-intensive and can only be justified where there are specific benefits. In some areas, young tea benefits from a grass mulch put down after planting. In other areas, such a practice has a negative effect.

A number of other materials have been tried as mulch in tea. Factory tea waste is usable in small quantities; a very deep layer can create problems. Any material which might be available should be tried initially on a small scale. Mwakha reported benefits from usage of cattle manure (Mwakha, 1990b) and sugar-cane filter-press cake (Mwakha, 1990a).

Some organic materials may be suitable for incorporation into planting holes. Sheep manure and chicken manure have both been used successfully. The latter is acidic and has been found to be useful where the soil pH is high. Such materials are likely to be more available and of greater benefit to smallholders.

FERTILIZER APPLICATION RATES

Every significant tea-growing area has a series of recommendations for fertilizers. In many cases these have arisen from research in local institutions. These are too many for individual listing in this volume.

Nitrogen applications vary and can be as high as 300–400 kg ha^{-1} per annum.

Phosphate applications do not normally go much higher than 120 kg P_2O_5 ha^{-1} per annum, but in most areas regular applications are made.

Potash applications vary greatly from area to area and also from one year to the next. Rates of application can be over 300 kg ha^{-1} per annum in the year of pruning. In many areas, 'remedial' applications are made in the year of pruning, in some cases based on the results of foliar analysis.

The above three nutrients are often applied as NPK mixtures or compounds. As examples, NPK 25:5:5 is a basic standard for several parts of Africa, north India recommends a series NPK 10:5:10, 10:2:4 and 12:4:8 for various ages of tea, and Sri Lanka recommends 10:1:4 and 4:1:3 for various ages. To all of these is often added extra potash fertilizer, particularly in the year of pruning. The formulae usually include some sulphur, which may be included as sulphate of potash or as elementary sulphur. Magnesium is also included in some formulations.

Foliar sprays of zinc are very common, usually zinc sulphate at about 20 kg ha^{-1} per annum spread over four applications per year.

In Asia, copper fungicides are used to control blister blight, so nutritional sprays of copper are unnecessary. Elsewhere, sprays of up to 20 kg ha^{-1} per annum of copper sulphate are given where it is found to be necessary.

DIAGNOSIS OF MINERAL DEFICIENCIES AND TOXICITIES

These become obvious when the characteristic symptoms appear. However, a lack of the nutrient concerned, to a level above that at which it causes deficiency symptoms, has probably restricted yield for some time before the symptoms appear. Chemical analysis of leaves can show such lesser deficiencies and enable corrective action to be taken before the situation is bad enough to damage the plants. This approach is now used in many industries in the world. In tea, it is not fully accepted in some areas. Others have found that it works well.

It is vital that the leaf sampled is described accurately and that all samples are the same. The state of the bush, age of leaf and other relevant conditions should be as consistent as is possible. Several organizations have published detailed criteria for content of various nutrients in tea leaves, which have been used successfully for diagnosing deficiencies and treating them, as well as helping towards understanding of results. Table 17.1 lists critical data used successfully by the author. In this case, the sample is the third leaf, when the two-and-a-bud flush above is ready for plucking. A minimum of 100 leaves from bushes of similar age is necessary.

FERTILIZERS AND TEA QUALITY

The quality of the made tea determines its sale price and is obviously of critical importance. Numerous investigations have been made to relate tea quality to fertilizers applied to the crop. No major effects have been reported. Willson and Choudhury (1969) found a small improvement in quality following the application of moderate quantities of phosphate fertilizer. This was confirmed by Dev Choudhury in 1993. There is a tendency for nitrogen to reduce the quality of black tea but improve green tea and for potassium to improve black tea. It can reasonably be concluded that fertilizer applications balanced to the needs of the plant for high yield will give a good-quality product. Argunova and Pritula (1991) studied the effects of nitrogen, phosphorus and potassium on tea quality. There are many other factors which can have a far greater impact on quality. Owuor (1993) reviewed factors affecting the quality of tea.

Table 17.1. Critical levels of nutrients in the third leaf of a tea shoot (from Willson, 1991).

	N (%)	P (%)	K (%)	Ca (%)	Mg (%)	Na (%)	S (%)	Fe (ppm)	Mn (ppm)	Zn (ppm)	Cu (ppm)	B (ppm)	Mo (ppm)	Cl (%)	Al (ppm)
Deficient	3.00	0.35	1.60	0.05	0.05	–	–	60	50	20	10	8	–	–	–
Subnormal	4.00	0.40	2.00	0.10	0	–	0.05	100	100	25	15	12	–	–	–
Normal	5.00	0.50	3.00	0.35	0.30	0.10	0.50	500	5000	50	30	100	–	0.10	–

N, nitrogen; P, phosphorus; K, potassium; Ca, calcium; Mg, magnesium; Na, sodium; Fe, iron; Mn, manganese; Zn, zinc; Cu, copper; B, boron; Mo, molybdenum; Cl, chlorine; Al, aluminium.

18

PESTS, DISEASES AND WEED CONTROL

Tea is native to Asia and some of the pests and diseases that attack the plant are not found elsewhere in the world. Many are, of course, universal and not specific to tea. Some originated in Asia and have been carried to other countries. Careful control and inspection and quarantine when a risk exists are essential when tea plants are moved from one country to another. The perennial monoculture of tea facilitates the continuous presence of pests and diseases and hinders control.

INTEGRATED MANAGEMENT

The management of a tea plantation must aim to minimize all the possible adverse factors, including pests, diseases and weeds. It is very difficult to achieve complete extinction of all such threats to high productivity within the plantation and its environs. Good management will keep them at a level within the plantation which is consistent with high productivity; failure to maintain the system will permit them to reinvade.

Pest management in major growing regions has been discussed by Mukherjee and Singh (1993) for North India and by Muraleedharan (1991) for South India. Control of diseases must also be a part of the management strategy.

Development of cultivars resistant to major pathogens should be a major part of plant improvement. So far, success in this objective has been limited.

Control in the field starts with land clearance. As far as possible, all pieces of old tree roots which might infect tea with *Armillaria* or other root diseases must be removed. Minimizing the loss of topsoil helps the tea plants to grow strongly, so that they will be more resistant to chance infection. The correct fertilizers in the planting hole are beneficial for the same reason; potassium is particularly important. A cover crop should be established as soon as possible; the species chosen must not be an alternative host for any disease or pest. The material planted out must be free of pests and diseases. Clones, or jats, to be planted should be chosen to be, as far as possible, resistant to local pests and diseases. Transport of tea plants must be controlled under quarantine regulations. Control of *Pratylenchus loosi* in Sri Lanka has led to prohibition of the movement of rooted cuttings from one estate to another.

Bushes must be managed so that a complete cover of foliage is achieved as soon as possible; the resulting low light level discourages weeds. Some weeding will inevitably be necessary, preferably by using herbicides, as implements can damage roots and stems, making entry points for pests and diseases. Prunings must be left on the ground to rot down to form a topsoil, which, by its origin, contains no weed seeds and no harmful spores, etc. Other field operations, such as fertilizer spreading, harvesting and shade-tree thinning, must be carried out in such a way that damage to the tea plants is minimized.

Land which has not been planted to tea and is close to the plantation should be kept free of alternative hosts for pests or diseases.

Biological control has been attempted for control of several pests. The only fully successful application so far is the control of *Homona coffearia* in Sri Lanka by the introduced parasite *Macrocentrus homonae*. There are several examples of partial control by biological agents; for example, *Phytodietus spinipes* limits the number of *H. coffearia* (Muraleedharan and Selvasundaram, 1991).

The use of pesticides should be kept to a minimum, as the balance of pests and predators can be disturbed. Some chemical treatments are of greater value than others – for example, fumigation of nursery soil. Local advice is available in most tea-growing areas, and chemicals used must be approved by the local government; Singh (1992b) listed the chemicals approved in India. Such lists change as new chemicals appear and older ones become prohibited.

PESTS

Cramer (1967c) estimated that the loss of potential crop from pest damage amounted to 8% in Asia, 5.3% in Africa and 4.7% in South America. These figures are significantly lower than those for coffee and cocoa. This may be, in part, a consequence of the product being foliage rather than fruit. Pests of flowers and fruit cannot affect the normal harvest. Over 300 species of animals have been reported as feeding on tea. Only a few of these create sufficient damage to justify preventive measures; the major pests are discussed in the following pages. Senaratne (1986) discussed the pests of tea in Sri Lanka.

Foliage Feeders

Helopeltis spp.

Mosquito bugs, *Helopeltis schoutedenii* and *Helopeltis amphila* in Africa and *Helopeltis theivora* in India, can create significant damage, causing crop loss in the field and affecting young plants in nurseries severely. They feed on young leaves, stems and buds. Leaves distort, blacken and die, so that bushes look scorched. Application of pesticides is necessary when insect counts are high and especially on young and pruned tea, which are particularly susceptible. Endosulfan or Ripcord are recommended in Kenya (TRFK, 1986g). Peregrine (1991) suggests that annatto (*Bixa orellano*) can reduce infestation by attracting the pest.

Thrips

South African citrus thrips, *Scirtothrips aurantii*, yellow tea thrips, *Scirtothrips kenyesis* and black tea thrips *Heliothrips haemorrhoidalis* are important pests in Africa. *Scirtothrips dorsalis* is the most common in Asia, while *Scirtothrips bispinosus* and *H. haemorrhoidalis* can also inflict severe damage. The insects feed on the young tender parts of the plant. Growing shoots are stunted, leaves are small and may fall off. Thrips are most active in unshaded tea in hot dry weather. Bushes are attacked most severely after pruning; the degree of damage can be minimized by careful choice of pruning date. Very severe attacks can be controlled by spraying endosulfan or fenitrothion (TRFK, 1986g).

Mites

In Africa, scarlet mites *Brevipalpus phoenicis* and *Brevipalpus obovatus* can cause significant losses: *B. phoenicis* is significant in Asia also. Yellow tea mite, *Polyphagotarsonemus latus*, is a serious pest in Africa. Purple mite, *Calacarus carinatus*, is important in Asia and Africa, attacking mature leaves, while other mites prefer young foliage. Among spider mites, *Oligonychus coffeae* is widespread in Asia, while *Tetranychus kanzawai* is common in Japan and Taiwan. *Acaphylla theae* is widespread in Asia. Mites mainly live on the underside of leaves, causing discoloration to a shade of brown which is dependent on the species involved. Mixed infestations occur.

Some clones resistant to spider mite have been identified in Kenya (Sudoi, 1990). Recommendations for insecticidal control include permethrin and propargite with nitrogen (Sudoi, 1990), dicofol, tetradifon and dimethoate (TRFK, 1986g), and the foregoing plus quinomethionate (Senaratne, 1986).

Cicadellids

Empoasca flavescens is widely distributed in Asia. Present throughout the year in north-east India, they are very numerous as 'green fly' in Darjeeling in June and July. *Empoasca onuki* and *Empoasca formosana* are present in Japan and Taiwan, respectively. These pests suck the sap from the undersides of young leaves, which curl downwards, brown and dry up.

Scale insects and mealybugs

These affect nurseries and mature fields. They are troublesome at times in Africa. In Asia, *Saissetia coffeae, Saissetia formicari, Eriochiton theae, Pinnaspis theae, Fiorinia theae, Phenascaspis manii* and *Pseudaonidia duplex* can all inflict serious damage. *Nipaecoccus viridis* is a common mealybug and causes defoliation and drying up in patches. Sooty moulds, such as *Capnodium* spp. and *Meliola* spp., often follow attacks of these insects.

Treatment with carbaryl or dimethoate plus white oil will give control; a second spray might be necessary. Attendant ants can be controlled by a ground spray of endosulfan (TRFK, 1986g).

Aphids

Toxoptera aurantii, is common, attacking tender shoots and leaves. Leaves curl and shoots are stunted.

Caterpillars

Many attack tea. In Africa, gelatine grub, *Niphadolepis alianta*, nettle grubs, *Parasa* spp., and tea tortrix, *Tortrix dinota*, attack sporadically. Tortrix species in Asia are *H. coffearia*, widespread everywhere except Japan, where *Homona magnanima* and *Adoxophyes* spp. are common. These caterpillars make nests by webbing between leaves. *Cydia leucostoma* and *Caloptilia theivora* are also important. *Cydia* (flushworm) larvae bore into and feed on the unopened bud. *Caloptilia* is a leaf miner and rolls up leaves. *Thosea cervina, Parasa lepida, Macroplectra natruria* and *Belippa lalaena* are less common. Feeding usually on older leaves, they are unpleasant, as workers are stung by the hairs. There are many other leaf feeders seen occasionally on tea.

Outbreaks of *H. coffearia* and *M. natruria* in Sri Lanka can be controlled by trichlorphon and methomyl (Senaratne, 1986). *H. coffearia* can be controlled biologically (p. 228). Neem cake has been found to inhibit caterpillars from feeding on leaves (Kakoty *et al.*, 1993).

Pests of Flowers and Seeds

Since vegetative propagation became the mainstay of production, the pests of seeds have become of minor importance. Where breeding for production of improved varieties is carried out and where tea-seed oil is produced, seed pests are important. In Africa, the larvae of the false codling moth, *Cryptophlebia leucotreta*, destroys the cotyledons. In Asia, *Poecilocoris latus* attacks young seeds, which fall prematurely. In mature seeds, only the pericarp is damaged. In China, a weevil, *Curculio styracis*, feeds on the contents of the seed.

Pests of Stems

Carpenter moths

In Africa, *Teregra quadrangula* larvae can ring-bark and kill young plants. Mature tea that has been attacked forms a callus over the wound. Young tea close to infested mature tea must be protected by insecticide spray of the mature tea, after pruning and before the larvae make a web in which they are protected.

Cockchafer

Schizonycha spp. sometimes damages stems and roots of young tea.

Cutworms

Agrostis spp. sometimes eat the stems of young plants in the nursery.

Stem borers

In Asia, several stem borers are serious pests on occasions. The shot-hole borer, *Euwallacea fornicatus* is a serious pest in South India and Sri Lanka, mainly below 1500 m altitude. Galleries are bored in stems of pencil thickness, which then break, causing a loss of crop. Red coffee borer, *Zeuzera coffeae*, often damages young plants. *Sahydrassis malabaricus* occasionally becomes a serious problem in South India. *Indarbela theivora* and *Indarbela quadrinotata* bore a small hole and spin a spiral run, covered by silk and powdered wood, around the branch. *Parametriotes theae* is a stem-boring caterpillar which is a serious pest in China, Georgia and surrounding regions.

Shot-hole borer damage can be minimized by careful choice of pruning time, reinforced by spraying with fenthion (Senaratne, 1986). Substantial reduction of infestations was achieved by spraying of fenvalerate, cypermethrin, oxydemeton-methyl or a mixture of quinalphos with dichlorvos (Muraleedharan *et al.*, 1992).

Termites

These are troublesome at times, particularly in dry weather. In Africa, *Macrotermes natalensis* and *Pseudocanthotermes militaris* sometimes ring-bark or sever young plants. The frames of mature bushes can be attacked. *Coptotermes havilandi* build galleries inside main stems, thus killing mature bushes in Mauritius. In Sri Lanka, *Postelectrotermes militaris*, *Neotermes greeni* and *Glyptotermes dilatatus* cause severe losses. *Microcerotermes* is the main pest in Bangladesh and North India. Dieback and rot following pruning facilitate the entry of livewood termites. Termites which only scavenge dead wood, such as *Odontermes* spp., are a minor nuisance.

In Sri Lanka, areas infested with termites are cleared and the area rehabilitated under Guatemala grass (*Trypsacum laxum*) for 3 years before replanting tea (Senaratne, 1986).

Pests of Roots

Nematodes

Over 40 species of parasitic nematodes have been reported in tea, but only a few are important. They cause sporadic damage outside Asia, particularly in nurseries. Subsoil used for cutting nurseries should be free of nematodes. Campos *et al.* (1990c) reviewed the nematodes which attack tea.

Of the root-knot nematodes, only *Meloidogyne brevicauda* attacks mature plants in Sri Lanka; the others only attack nursery plants. *Meloidogyne incognita* can appear in nurseries worldwide; *Meloidogyne javanica* is more common in Asia.

The lesion nematode, *P. loosi*, is a serious pest in Sri Lanka, and in China and Japan to a lesser extent. Leaves on affected plants become yellow and smaller; affected plants can die in drought or after pruning.

The burrowing nematode, *Radopholus similis*, has become a serious pest in Sri Lanka.

Helicotylenchus spp. reduce tea productivity in South India. The sheath nematode, *Hemimicronemoides kanayensis*, harms deep feeder roots in Japan and Taiwan.

Problems with nematodes in nurseries can be avoided by fumigating all nursery soil. When replanting tea fields, care must be taken to remove all tea roots, which could be infested. Resistant clones are available for replanting, but there is no other way to deal with infested mature tea.

Insects

Some insects feed on roots; white grubs, *Helotrichia* spp., are the most important. They feed off well-formed roots and stem bark just below ground level.

DISEASES

Most diseases of tea are fungal; bacteria, algae and viruses cause but a few. Losses of the potential production due to diseases are estimated by Cramer (1967c) to be 16% in Asia, 3.9% in Africa and 3% in South America.

Diseases Affecting Leaves

Blister blight (Fig. 18.1)

Exobasidium vexans is a major factor in tea-growing in Asia. Known for 125 years in North India, it did not appear in South India and Sri Lanka until 1946 and Indonesia in 1950. This disease only affects succulent young leaves and stems, mature leaves and stems are not affected.

Pale yellow, translucent spots appear on young leaves about 9 days after

Fig. 18.1. Tea blister blight: blisters on leaves (photograph from J.M. Waller, CABI Bioscience).

infection. These develop into blisters on the underside, with indentations on the upper side, after about a further 9 days. After about 3 more days, sporulation turns the blisters white. Then they turn brown and several blisters on a leaf make a leaf distort. Affected young stems become bent and distorted and may break and die. This is a severe setback, particularly when recovering from pruning. Repeated attacks debilitate and kill bushes. The disease is more prevalent at higher altitudes and during the wet season. Sugha *et al.* (1990) discussed factors affecting the development of blister blight. To minimize the incidence of the disease in Himachal Pradesh, India, tea should be unshaded and pruning in June avoided.

Copper-based fungicides give effective control. The application of 280 g ha^{-1} of a 50% copper formulation every 7–10 days is normally sufficient. More frequent and concentrated applications must be made on tea recovering from pruning. Sugha (1991) investigated the comparative efficiency of fungicides. Some systemic fungicides have been found to give effective control and are being used on an increasing scale. Updated control recommendations have been published by Mouli (1993). Sugha *et al.* (1991) investigated the effect of blister blight on tea quality.

Net blister blight

Exobasidium reticulum is also only known in Japan and Taiwan. White reticulated lesions appear on veins or the underside of leaves. If this disease appears in the autumn, leaf fall and stem dieback will hinder the regrowth necessary for the next year's production. Copper-based fungicides or chlorothanil will control this disease.

Anthracnose

Colletotrichum theae-sinensis is important in Japan and Taiwan but not known elsewhere. It forms large reddish-brown lesions on leaves, most of which later fall off. It is more prevalent at high altitudes and in wet conditions. Outbreaks in autumn adversely affect the growth of new stems for the following year's production. The disease can be controlled by spraying fungicides based on copper, chlorothalonil or thiophanate.

Colletotrichum camilleae

See brown blight below.

Eye spot

Pseudocercospora ocellata is common in Central Africa in the wet season but is not a serious problem.

Calonectria

Calonectria spp. cause brown spots on leaves in Mauritius. The leaves rot and go black. Copper-based fungicides will control serious outbreaks.

Brown and grey blights

Brown blight, *Glomerella cingulata* syn. *C. camilleae*, and grey blight, *Pestalotiopsis theae*, are weak pathogens which occur on bushes weakened by poor nutrition, damage from implements, plucking, herbicides, hail, waterlogging or extremely low temperatures. Brown blight occurs on leaves of all ages and can cause defoliation. Grey blight may cause sporadic attacks on nursery plants but more usually appears on old leaves. These diseases are known in most tea areas. In Japan, grey blight is caused by *Pestalotiopsis longisepta* and is important because it stimulates *G. cingulata*, which causes severe damage on mechanically harvested tea. Thiophanate-methyl applied 3–5 days after harvesting controls these outbreaks. Elsewhere, these diseases will disappear when the conditions weakening the bushes are rectified.

Bacterial shoot blight

Pseudomonas theae kills leaves and tips of shoots in Asia when plants are less active at the start and end of the season. Damage is worse in young tea and in windy situations. It can be controlled by copper, as Bordeaux mixture.

Red rust

Red rust, *Cephaleuros parasiticus*, is caused by an alga and attacks tea frequently in low-altitude areas of North India and Sri Lanka. It has a more severe effect on weak bushes, although it is frequently seen on healthy plants.

Leaves on affected stems become variegated – green with either yellow or white – and may die back. Infected areas develop red or orange patches when the alga produces fructifications, normally between April and July. The disease will weaken bushes progressively. Copper-based fungicides will control the disease. Some resistance to this disease has been found among clones in North India (Singh, 1992a).

Stem Diseases

The regular pruning of tea leaves large surfaces exposed. These provide entry points for diseases. Sunscorch following pruning can increase the area open to attack.

Cankers

Stem and branch canker, *Phomopsis theae*, is known in Africa and Asia. It mainly affects young plants of susceptible clones. Cankers arise on the collar and on stems. Collar cankers are more serious but rarely kill bushes. They can provide entry points for termites. The incidence of this disease is related to water stress, so management to minimize this should reduce the frequency of attacks.

Wood Rot

Wood rot disease, *Hypoxylon serpens* and *Hypoxylon investiens*, mostly affects older tea. Affected stems become dry, light, brittle and inactive. Characteristic fructifications are formed: irregular, slightly raised, whitish-grey to dark patches. This disease is important in East and Central Africa and *H. serpens* occurs in North India.

An effective control has not yet been devised. Copper oxychloride in linseed oil painted on pruning cuts has been suggested in Kenya.

Pink disease

Corticium salmonicolor affects branches occasionally in Equatorial Africa. The bark cracks and is covered by pinkish-white mycelium. Affected branches should be removed and the plant treated with copper oxychloride. *Pellicularia salmonicolor* causes dieback of young shoots in Asia. The bark of stems is killed in patches, where the wood later rots to leave a hollow channel.

Minor cankers

Nectria cinnabarina produces cankers where stems have been damaged. Affected stems die back. It is known mainly in North India and occasionally in Sri Lanka. *Hypoxylon asarcodes* kills bark in longitudinal patches. Fructification produces patches like dry black paint. This fungus causes tarry root rot. *Leptothyrium theae* causes stem cankers which can kill young tea. *Macrophoma theicola* causes stem cankers and twig dieback. This disease may kill young stems or occasionally main stems by girdling. Branch canker, *Poria hypobrunnea*, is prevalent in North India. The fungus enters by a wound and spreads until it kills the whole plant.

Thorny stem blight, *Aglaospora aculeata*, is common in Darjeeling and Sri Lanka. The fungus enters through wounds and spreads downwards, killing the whole plant. The wood of the stem shows black patches, with mycelium on the surface. The characteristic black thorny protuberances on dead stems are fruiting bodies of the pathogen. There is no cure for this disease, but it is a disease of weak bushes, so good management should minimize its incidence.

Root Diseases

Armillaria mellea

Honey fungus, root splitting disease, *Armillaria mellea*, occurs throughout the world. It causes significant losses in Africa, is common in Indonesia but is rare in tea in North India and does not affect tea in Sri Lanka. It usually spreads from dead roots of old forest trees, shade trees or previous crop trees. It invades the tea root system and develops a thick layer of mycelia between the bark and wood. The collar and roots often show longitudinal cracks. The main preventive measure is to ensure that all trees in tea fields, either former forest trees or superfluous shade trees, are thoroughly killed before planting tea. Ring-barking ensures a slow death, which exhausts *Armillaria* if present. All tree roots must be removed when clearing land. Onsando and Waudo (1994) reported that several *Trichodeoma* species inhibited the growth of *Armillaria*.

Charcoal stump rot

Ustulina deusta is sporadic in Africa but widespread in Asia. Under the bark, the wood is covered by fan-shaped mycelium patches. Black lines or bands are seen in the wood. The fructifications are ovoid and furrowed; when old, they are black like charcoal. It spreads by spores and by contact. Killing by ring-barking and removal of unwanted trees minimizes its incidence.

Red root disease

Poria hypolateritia is common and economically important in Asia. Mycelium can be seen on the root surface. Initially white, it compacts into dark red cords or sheets. It spreads by mycelial strands from infected stumps. Diseased areas (not planted with tea) can be cleaned by methyl bromide fumigation or with systemic fungicides.

Brown root disease

Fomes noxius mycelia bind soil around the root. Mycelium is usually found under the bark also. The wood becomes permeated with yellow-brown mycelium. This disease also spreads from diseased stumps.

Violet root disease

Sphaerostilbe repens is widespread in Assam but rare elsewhere. It develops in waterlogged conditions. Roots of affected plants are black or light violet in colour and have a characteristic odour. The wood under the bark is covered with thick purple-black strands.

Other root diseases occur occasionally in various areas.

MANAGEMENT OF DISEASES

The incidence of many diseases can be minimized by good management. Correct nutrition, good drainage, avoidance of damage and removal of all old tree material are essential. Barua (1991) reviewed cultural measures for the cold season which reduced disease incidence. The severe leaf diseases, such as blister blight, can only be controlled by fungicides. Even with these, correct nutrition will help to minimize the severity of attacks.

Most diseases affecting stems and leaves are controlled by copper or compounds containing copper. The inorganic copper formulations are not toxic, making them easier and safer to apply. The copper is absorbed by the plant, so these treatments also automatically provide a protection against copper deficiency.

Organic systemic fungicides are effective against some diseases and are used as alternatives to copper. They can be applied by modern ultra-low volume spraying techniques. These reduce the amount of water needed by a substantial amount, thus saving most of the effort involved in carrying water to the field. They do need more care in handling.

Although there is a range of characters among the tea plants in plantations, it has not yet been possible to isolate or breed plants with total or a high degree of resistance to diseases. Clones with some resistance have been planted where disease problems are serious, in order to minimize losses.

WEED CONTROL

Cramer (1967c) estimated that the losses of crop below the potential maximum due to weeds are 9% in Asia, 6.6% in Africa and 6.8% in South America. With high-quality management, losses of tea crop due to weeds should be insignificant.

Weeds are a minor problem in mature tea with a full cover of tea and a deep mulch from prunings and leaf litter below. Regular weeding should not be necessary, although some work may be necessary after pruning. It is vital to keep disturbance of the mulch and topsoil to a minimum – preferably none at all.

Weed growth immediately after planting should be minimized by the cover crop. Such weeds as do grow can be controlled by contact herbicide (glyphosate or paraquat). Care must be taken not to allow the herbicide to get on to the young tea plants. Hand-weeding will have to be done carefully around these. As the tea grows and the tea bushes get larger, their shade will reduce weed growth. The mulch will begin to form and enhance the effect. As soon as the tea plants have little or no foliage close to the ground, blanket or spot spraying of the contact herbicide will control weeds. The amount of this necessary will recede as the tea grows. Field margins, however, will continue to need spray treatment at intervals.

If tea has become infested by weeds, large quantities of herbicide will be necessary for an initial period. Frequent treatments by glyphosate or paraquat will clear most weeds. If couch grass is present, applications of dalapon will kill it. Large, deep-rooted weeds must be pulled or dug out. Diuron is a persistent herbicide; sprayed carefully on the ground without touching the tea, it will prevent most weeds from regrowing for several weeks. Kabir *et al.* (1991) evaluated the effectiveness and effects on tea yields of five herbicides in Darjeeling.

The establishment of a deep surface mulch from prunings and leaf fall and bush management to ensure complete ground cover, giving a minimum of light at ground level, will greatly minimize, if not completely prevent, weed growth. Without the surface mulch and ground cover, weed control will be difficult and expensive. It may be necessary on occasions to spray a contact herbicide when weeds have appeared after pruning. The benefits of chemical weed control in terms of tea yield were demonstrated by Rahman (1991).

19

Crop Processing at the Plantation and for Retail Sale

GENERAL

All three crops must be processed to some degree at the plantation. As harvested, none of them is in a condition in which it could avoid rapid deterioration during transportation. It has also been found that plantation processing has an important effect on the quality, i.e. the aroma and flavour, of the final product. As the price paid for the product by the consumer is related to quality, it is important for the plantation to despatch its product at the highest possible quality.

At the very least, all the products need drying; as harvested, they contain a high proportion of water. Without drying, they will deteriorate quickly. Drying can take place at different stages of processing – as a first or last operation or at an intermediate stage. It also happens that, in the processing of each of the crops, there is a stage called 'fermentation'. This is not necessarily a true fermentation, i.e. a chemical change carried out by a yeast, but instead a stage at which chemical changes occur which are vital for the development of the final quality of the produce.

All three crops go through other processes, which do not necessarily have to be done in the country of origin, before they are put on retail sale. Traditionally, some of these operations were carried out in the consuming country, but there is an increasing trend for these processes to be located in producing countries, thereby providing more added value and therefore revenue. Employment opportunities for the local population are also created.

COFFEE

Processing at the Plantation

Figure 4.6 (p. 38) shows a cross-section of a coffee fruit ('berry' or 'cherry'). The various layers which surround the bean (cotyledon) are indicated. The presence of the mesocarp layer (the 'pulp' or 'mucilage') means that pressure on one end of the cherry will eject the bean from the mesocarp and exocarp. The final product required from the plantation is the bean, stripped of all outer layers and dried to a moisture level at which it will keep without short-term deterioration.

Coffee-processing operations at plantations can be divided into 'wet' and 'dry' processes.

Wet processing

This process requires cherries which are fully ripe or very close thereto. Under-ripe and overripe cherries give coffee with a poorer quality; hard, dried, overripe ones are rejected by the pulper. The cherries must be processed as soon as possible after harvesting; wet fruit start to degrade as soon as they are removed from the tree.

The cherries are moved through the process by a flow of water. An adequate and reliable source of water is therefore required. It is convenient to site the processing factory on a hillside so that a flow of water under gravity moves the coffee through the process.

Harvested cherries are put into a separating tank full of water. Stones and other heavy impurities are removed at the bottom; hard, partially dried cherries float and are discarded from the top. Good cherries sink, are withdrawn in a stream of water at an appropriate level above the tank bottom and are taken to the pulper. The pulper consists of a drum or a series of discs, which carry metal plates with teeth on the exposed surfaces. Stationary knives are fixed close to the teeth. As the rotors revolve, cherries are pulled by the teeth against the knives, and the pressure generated expels the beans from the cherry. The pulp (mesocarp) and skin (exocarp) go through the gap and are discharged at the rear, while the beans pass over the knives and fall at the front on to a screen. Cleaned beans fall through the screen, beans with pulp still attached go over the screen to a second pulper, known as the 'repasser'. At this stage, the beans carry a residual film of mucilage.

The pulp separated from the beans is a waste product, which may merely be dumped. This can lead to environmental problems; pulp in quantity will foul watercourses. If it is returned to the plantation, its nutrient content is recycled to the coffee. Pulp is difficult to handle, as it has a high content of water (75–85%); pumping is the best way to move it, but a suitable pump and piping have to be installed. Braham and Bressani (1979) discussed possible uses for coffee pulp and treatments which are necessary before it can be used; treatments discussed include drying, ensilage and reaction with various chemicals. It is suggested that incorporation of treated pulp in animal feed is the use with the greatest potential. Depending on the animal to be fed and the treatment process used, approximately 15–20% of the normal feed can be replaced by coffee pulp, releasing other ingredients, such as maize and soybean, which can be used by humans. The use of coffee waste in biogas production has also been investigated (Boopathy, 1989), with respect to design, potential output and economics. The waste water from the factory is also noxious and will foul watercourses; restrictions on discharging this water are being drafted in several countries. Castillo *et al.* (1993) reported a project for treatment of this waste water.

The beans are carried forward by the stream of water into a channel, where they are washed by the flow of water, and then continue into tanks for 'fermentation', where they accumulate to a depth of about 1 m (Fig. 19.1). The so-called 'fermentation' process is an enzymic reaction which decomposes the mucilage, and it can be accelerated by addition of a pectic enzyme. It may be carried out 'dry', i.e. with no added water and the drain open, or 'wet', i.e. with the beans submerged in water in the tank, followed by a short dry fermentation. The beans are then washed to remove decomposition products, in long channels or machines, and dried. Soaking the beans during wet fermentation and washing improves quality by removing brown oxidized polyphenols to leave the beans with a desirable blue colour.

Drying is carried out either in the sun or by warm air in machines or partly by each. Excessive heat lowers quality. Some sun-drying helps to produce a good quality, as sunlight causes photochemical changes during drying from 30% to 20% moisture which improve the colour and flavour of the beans.

Fig. 19.1. Coffee-fermenting tanks. Levelling the beans which have been run into the tanks by the stream of water from the pulper. (Photograph from the International Coffee Organization, London (ICO).)

The beans are spread out either on a concrete floor or on raised mesh benches (Fig. 19.2). Covers must be available to protect drying coffee overnight or during rain. Coffee which is drying in the sun must be turned frequently.

The dried beans still carry the endocarp (hull or parchment). At this stage, they are known as 'parchment coffee'. The hull is removed by machine in a process called 'hulling' or 'milling'; this is rarely carried out on an estate – the parchment coffee is usually despatched to a central factory for milling. Finally, a polishing process removes the last of the silverskin; grading and packing for export follow. In its final stage, the coffee is referred to as 'green beans', as good-quality beans have a blue-green colour.

Disposal of waste from hulling and polishing is a problem. Coffee husk is used as fuel for coffee driers, for making charcoal briquettes or as mulch in coffee fields. Burning coffee husk to dry the beans makes a significant improvement in profitability, particularly if drying is by a fluidized bed. Yoshioka *et al.* (1992) reviewed the production of liquid organic chemicals from various waste materials, including that from coffee processing.

Dry processing

With arabica coffee in Brazil and in a high proportion of robusta coffee, the cherry is stripped from the trees when a large proportion is ripe. Some 'classification' may follow, i.e. removal of underripe cherries by washing in running

Fig. 19.2. Spreading coffee beans on a raised mesh bench for drying (photograph from the International Coffee Organization, London (ICO)).

water and screening. This process will also remove adventitious non-coffee rubbish. The washed cherries may still include some underripe and overripe cherries. They are spread out to dry, usually on a concrete floor, known as a 'barbecue', in the sun. Sometimes beans are spread on mats raised above ground. Machines are used in some places to complete drying.

The dried cherries have then to be dehusked, which is much the same as the hulling carried out on dry parchment coffee from the wet process. Similar machines are used, but they have to handle the much larger amount of material – dried mesocarp and exocarp – removed from the beans. This is followed by polishing.

Polishing

This process is particularly important for dry-process robusta, where the silverskin is firmly attached. Water may be added to the polishing to give 'washed and cleaned' beans.

Coffee quality

The quality of coffee is a combination of flavour and aroma. These come from the chemical constituents of the coffee beans. Chemical reactions take place during processing, particularly in the fermentation stage of wet processing, which produce the desired chemicals. The quality of the cherries influences the quality of the product; underripe, overripe, diseased and damaged cherries reduce quality, as do cherries that have been delayed between harvesting and processing. The final assessment of quality is made subjectively by skilled tasters, who consider appearance, flavour and aroma. The quality so assessed has a marked influence on the price at which the coffee can be sold.

In general, dry-process coffee is regarded as of lower quality than wet-process coffee. There are exceptions, such as the valued 'mocha'-type coffees from Brazil. Robusta coffee is of lower quality than arabica.

Barel and Jacquet (1994) reviewed the factors that control coffee quality.

Grading

Coffee beans are graded primarily by size by passing them over a series of sieves. The size standards are set by commercial practice and contracts and an International Standard.

A number of other operations are often carried out to remove below-standard beans:

- Airlifting: a blast of air removes low-density beans.
- Electronic sorting identifies and removes discoloured beans.
- Ultraviolet excitation enables some malflavoured beans to be identified and removed.
- Manual sorting: skilled operatives can identify and remove low-quality beans.

Processing for Retail Sales

Blending

The green coffee beans are shipped to the consuming countries, where the final operations before retail sale are carried out. Blends may be made from several shipments of different origins; some may include both arabica and robusta.

Roasting

The blended beans are then roasted; this process develops the final flavour and aroma. The beans are heated and must be kept in motion during the process. Variations of temperature and duration of roasting affect the flavour of the final beverage. The flavour required depends on the user's preference, and is achieved by the choice of green beans, blending and roasting conditions. Small and Horrell (1993) reported the technology which produces roasted coffee of a lower density but higher flavour content, so that a given quantity of unroasted beans will make a greater volume of beverage.

COCOA

Processing on the Plantation

Opening pods

Pods have to be opened and the beans removed. This must be done in or close to the plantation. The opening of pods has been discussed in Chapter 11, p. 133.

Underripe beans do not make good cocoa, so pods should be opened when ripe. Pods can be kept for a few days between harvesting and opening. Such a delay has been found to be advantageous; the fermentation heats up more quickly and the proportion of undesirable 'purple beans' falls with delays of up to 4 days.

Fermentation

Beans must be fermented as soon as they are removed from the pod. Fermentation has four objectives.

1. To remove the mucilage attached to the beans.
2. To kill the embryo so that beans cannot germinate.
3. To encourage chemical changes within the bean which produce the substances responsible for the chocolate flavour.
4. To reduce the moisture content of the beans.

The conversion of the sugars in the mucilage to alcohol and then acetic acid is a true fermentation. A solution of acetic acid in water is formed, which drains off the beans. The chemical reaction raises the temperature and this, together with the acid drainings, kills the embryo. The other chemical changes occur within the beans.

Fermentation is carried out in one of two ways. Traditionally, the beans are heaped on to banana leaves on a bed of twigs and the finished heap covered with more banana leaves. The other method uses a series of rectangular wooden boxes with drainage holes to permit the acid liquor to escape (Fig. 19.3). Full boxes should be covered with banana leaves.

The size of heaps or boxes is limited by the need to ensure a sufficiently high temperature and to permit liquid to drain out and air to circulate freely around the beans. Small quantities, below about 70 kg, will not reach a sufficiently high temperature, while over about 150 kg aeration becomes restricted.

The heaps have to be turned over at intervals of 2 days so that all beans ferment equally. The temperature of the beans rises sharply to 40–45°C after the heap is made. After turning, the temperature will rise again, to 45–50°C. Fermentation usually takes 6–8 days for Forastero and 3–5 days for Criollo cocoa. Fermenting boxes usually have removable boards at the front, which makes them easy to empty. Boxes arranged like steps on a stand or hillside simplify turning by making it easy to transfer beans from one box to the next by gravity.

Fig. 19.3. Cocoa-fermenting boxes (photograph from W.R. Carpenter & Co. Ltd, Papua New Guinea).

The end of the fermentation has to be judged by experience. Beans in the corners of a box may be overfermented and have become almost black, with a smell of ammonia when the fermentation is finished. Most beans will be brown at this stage; if opened the cotyledons will be seen to be pale in the centre, with a brown ring.

Effendi (1993) reported the effects on cocoa flavour of storage before fermentation and the depth of fermenting boxes. Bhumibhamon *et al.* (1993) studied the microbiological, physical and chemical changes during fermentation.

Drying

Fermented beans must be dried to prevent deterioration. This is done mainly by spreading them out in the sun on concrete floors or on mats raised above ground. The beans need to be covered overnight and in rain. Sun-drying alone will take at least a week. Foreign matter can be picked out from the beans while they are spread out (Fig. 19.4).

Sun-drying can be supplemented by drying with hot air; a range of types of drier of varying complexity have been devised. Cunha (1991) describes the drier used traditionally in the state of Bahia, in Brazil, and a recent improvement incorporating a locally made fan. Grandegger (1989) described

Fig. 19.4. Cocoa-drying benches. Note covers on left which are moved over the beans at night and during rain. (Photograph from Dr K. Hardwick, Cocoa Research Unit.)

a solar-powered tunnel drier developed in Indonesia. The dried beans should have a moisture content of 6–7%; over 8% the beans can become mouldy and below 5% they are brittle. Dried beans are bagged for shipment to chocolate factories. Faborode (1991) investigated the drying of cocoa beans and the effects of rewetting partly dried beans; the physical and chemical properties were changed. Development of an artificial drier that simulates sun-drying, with its nightly rest period, is discussed. Solar-heated driers have given satisfactory results under test in Papua New Guinea.

Cocoa quality

The quality of cocoa determines the demand for the product and hence the price at which it can be sold; a low-grade product may be difficult to sell. Flavourwise, there are two types of cocoa: 'bulk' cocoa includes about 95% of world production, while 'fine' or 'flavour' cocoas, about 5% of world production, have special flavours which are required for dark (plain) chocolates and high-quality coatings; these products attract premium prices. Traditionally, fine cocoas come from Criollo or Trinitario trees and bulk cocoa from Forastero trees, but there are exceptions to this rule. The standard for bulk cocoa is based on the Ghanaian product. Fowler (1994) discussed the types and characteristics of fine cocoas, whilst Clapperton (1994) reviewed research on the factors affecting the quality of cocoa.

Stored beans attract insect pests: Sivapragasam *et al.* (1992) reported the insects invading bean stocks in Malaysia. Most of the invading insects in all their forms were killed by fumigation with methyl bromide, but some larvae survived.

Preparation of Cocoa and Chocolate for Retail Sale

Cleaning

The first process after beans have been received at the chocolate factory is cleaning. The beans are passed over a vibrating sieve, through which air is blown. Unwanted material either falls through the sieve or is blown away by the current of air. Magnets trap pieces of iron.

Roasting

The beans are roasted to develop the chocolate flavour. The exact conditions of temperature and duration of heating are adjusted to suit the beans being processed and the flavour required. The temperature will be between 100 and 150°C and the duration from 20 to 49 min. The process operates continuously and the beans are kept in motion to prevent local overheating. The beans are quickly cooled after roasting.

At present, almost the entire world output of cocoa is roasted in developed countries. However, Muhammad Nur Ahmad (1992) described an integrated roaster–sheller for use in rural (plantation) locations. Ghana has developed an industry that exports semifinished cocoa products.

Kibbling and winnowing

Kibbling is a means of breaking up the beans. The shell (testa) is broken into small pieces and blown away by a current of air (winnowing). The ensuing passage over vibrating screens separates the cotyledon material from any sprouts. The broken cotyledons are now called the 'nib'. Where necessary, the nibs from different origins are now blended together.

Cocoa shell is now used as a mulch in horticulture, partially replacing peat. It is therefore environmentally superior, as it is a totally waste product replacing peat, the extraction of which is now considered environmentally undesirable.

Grinding

The nib is ground to a fine powder at a temperature of 50–70°C. The product is known as the 'mass' or 'liquor'; it is liquid at the elevated temperature. At this stage the mass will contain from 55 to 58% of fat (cocoa butter).

Extraction of cocoa butter

For preparing cocoa for use in making a beverage, the fat content must be reduced. Hydraulic presses reduce the fat content to a little over 20%, an extrusion press will take it down to about 10% while solvent extraction will reduce it further.

Cocoa-powder finishing

The defatted nib is ground and passed through a fine sieve. It is common practice to treat the powder with alkali to control colour and make it disperse in water easily, before final drying and packing.

Chocolate manufacture

To produce chocolate, nib is mixed with sugar and cocoa butter to produce plain chocolate. For milk chocolate, milk is added to the mix.

The mixture is next ground between rollers and 'conched'. The latter process is a grinding under a roller while heating to reduce acidity by driving off fatty acids. The final product is completely homogeneous and smooth to the palate.

TEA

There are three types of dried tea, which differ in the degree to which they have been fermented.

1. Green tea, which is not fermented at all.
2. Semifermented or 'oolong' tea, which has been fermented for a relatively short time.
3. Black tea, which has been fermented to the extent that the dried leaf is black.

Green Tea

Processing on the plantation

Camellia sinensis var. *sinensis*, the China type of tea, must be used for green-tea production. The Assam type has too high a content of flavanols, which would make green tea taste excessively bitter.

The production of green tea is characterized by an initial heating process, which kills the enzyme, polyphenol oxidase, responsible for conversion of the flavanols in the leaf to the dark polyphenolic compounds that colour black tea. The other important process is rolling, which cuts and twists the leaves. The final form of green tea depends on the particular variant being produced. The rolling stage is very similar to the operation with the same name in black-tea production. In the main, green-tea production is restricted to China and Japan.

There are two main heating regimes used in green-tea production: steaming and pan-firing. Young leaves from tea grown under artificial shade, which is progressively darkened to intercept up to 98% of the light, are steamed to give the special 'Ceremony Tea'. These leaves have a low flavanol but high amino acid content when plucked after 20 days in the heavy shade. The leaves are steamed immediately after plucking and then dried in a tunnel drier. They are then pulverized in a stone mill, which produces the fine fluffy powder which is Ceremony Tea. The brew from this is green in colour and has a sweet taste.

Leaves from the unshaded tea that forms the bulk of Chinese and Japanese tea are also steamed. The leaves are harvested three times a year. Plucking is not restricted to the two, or three, leaves and a bud, which is the usual crop for black-tea production. Infrequent harvesting of young and old leaves gives very high yields (8000 kg ha^{-1} in spring, 6000 kg ha^{-1} in high summer and 4000 kg ha^{-1} in late summer). Very heavy applications of fertilizer are made to support these yields.

To prepare the tea, the leaves are steamed for 45–60 s. They are then curled (rolled) in hot air at 90–110°C for 40–50 min. The moisture content is reduced from 76% to about 50% during these operations. Fifteen minutes' rolling without heat follows and then 30–40 min drying in hot air at 50–60°C to reduce the moisture content to about 30%. The leaves are then rolled again on a pan at 80–90°C for 40 min. A final drying at 80°C continues until the moisture content is below 6%. The tea is now in the form of fine needles and is sieved to remove stalks and dust. A final drying then removes the green odour.

For pan-fired tea, the leaves are heated in a pan to 250–300°C for 10–15 min. The roller is designed to form the tea into the characteristic shapes, while agitation prevents the tea from burning. A final 10–15 min drying at 100–150°C completes the process. The final tea can be in a ball ('gun' or 'gunpowder' type), a fine twisted form or pale, white, polished, flat leaves. Some of the teas have special aromas, such as roast and smoky.

Semifermented or Oolong Tea

Plants of var. *sinensis* are used for this type of tea but the favoured clones have a wide range of leaf characters.

Fresh shoots are 'withered' by spreading out on bamboo mats for 30–60 min in sunlight. The withering then continues on a floor in a room for 6–8 h with gentle agitation once an hour. The leaves ferment during withering and develop a red colour. The fermentation is stopped by heating the leaves in a pan at 250–300°C for 15 min. The leaves are then rolled and finally dried.

Oolong tea was known in Europe as Bohea tea in the 18th century and initiated the taste for black tea, which became the usual type when Indian tea became available.

Black Tea

Tea quality

Black tea is sold according to its quality. There is a substantial price difference between high-quality and low-quality tea. Quality is controlled by several factors.

1. The tea bushes from which the leaves were plucked. Some clones and some seed sources ('jats') make better tea than others. China-type plants make poor black tea because the leaves have a lower content of flavanols than Assam-type tea.
2. The climate. Tea grown in a warm, tropical, lowland area grows very quickly but the leaves are of poor quality. In contrast, leaves from tea grown in tropical highlands are usually of high quality. Where the climate is such that there is a season of dormancy, leaves of very high quality are formed as a first regrowth after the first plucking when the dormancy breaks. The 'second flush' of Assam tea is renowned for its quality.
3. The leaves plucked from the bushes. The content of flavanols is highest in the bud, is lower in the first leaf and continues decreasing as the leaves age. The stalk has an even lower content. Therefore tea made from the buds only will have a very high quality; this was done in the past and the product known

as 'tips'. The yield was very small and the practice has long since been discontinued. Nevertheless, the word 'tips' indicates high quality and it is still used today, applied to good-quality brands.

Plucking two leaves and a bud (fine plucking) should produce a better tea than three leaves and a bud (coarse plucking). Plucking inevitably includes some stalk but this is mostly eliminated in the manufacturing process.

4. The design and control of the manufacturing process, to be described in the following pages.

Other factors may have small but significant effects on quality. The effects of fertilizer nutrients have been reported on a number of occasions. For most nutrients, the reported correlations between nutrient application and quality have been variable. The only correlation that has been reported consistently is a small improvement in quality following application of phosphate fertilizer in moderate amounts. First reported by Willson and Choudhury (1969), it was confirmed by Dev Choudhury in 1993.

Owuor (1993) reviewed the factors affecting the quality of tea.

The manufacturing process on the plantation

This includes the following stages, each of which will be described separately.

1. Withering.
2. 'Rolling'.
3. Fermentation.
4. Drying.
5. Sorting.
6. Packing.

Withering

This reduces the moisture content of the leaves from 75–80% to 55–70%. At the same time, chemical changes occur in the leaves which are an essential part of the development of the flavour of the tea.

This operation was carried out at one time by spreading the leaves thinly on shelves made of netting, called 'tats', which requires a large shelf area and hence a large building (Fig. 19.5). It is labour-intensive and not easily controlled; it would be uneconomic to duct and control the air flow to such a large area. Leaves are now spread on troughs. Made of steel, these are about 1 m deep and up to 2 m wide. The leaves rest on wire mesh near the top and are spread in a much thicker layer than on tats. A reversible fan at one end can blow air in either direction through the bed of leaves, which is usually about 200 mm deep. A method of warming the air is essential (Fig. 19.6).

The withering process takes several hours. Usually, tea leaves harvested one day are withered overnight and processed the following day. Withering must therefore be controlled so that it is complete when the leaves are due to be processed. If the process goes on for too long, the leaves become

Fig. 19.5. Large four-storey building to house 'tats' (netting shelves) for withering tea leaves. Now out of use, replaced by troughs (see Fig. 19.6). (Photograph from George Williamson & Co. Ltd, London.)

Fig. 19.6. Tea leaves on withering trough (photograph from George Williamson & Co. Ltd, London).

'overwithered', which will reduce the quality of the final product. The amount of moisture on the leaves and the humidity of the ambient air will vary with the climate; it is therefore essential to adjust the rate of withering by adjusting the temperature and flow rate of the air.

Rolling

This process has two objectives: to reduce the leaves into small pieces and to disrupt the leaf structure, so that the polyphenol oxidase is brought into contact with the flavanols and the fermentation can proceed efficiently. A variety of machines have been designed for this purpose.

- The roller. This machine was devised by William Jackson *c.* 1873 (McTear, 1958) and has become the traditional way of disrupting leaves; it has not yet been completely displaced by modern continuous machines. The withered leaves are loaded into a central vertical cylinder and put under a variable pressure by a piston that moves up and down the cylinder. Below the cylinder is a horizontal circular pan of larger diameter than the cylinder, with a corrugated upper surface. The leaves are compressed in the cylinder between the pan and the piston. The pan and cylinder each move horizontally in different circular orbits so that the leaves are torn and curled by passing over the corrugated surface of the pan. When the process is finished the leaves are discharged through a door in the pan (Fig. 19.7).

Fig. 19.7. Rollers for tea processing; note the vertical cylinder and horizontal pan. (Photograph from George Williamson & Co. Ltd, London.)

Tea manufacture largely or wholly by rollers is known as 'orthodox' manufacture. The leaves are loaded into rollers, discharged and reloaded several times before fermentation. The rolled leaves consist mainly of pieces up to about 4 mm in size and curled up. Rollers can only be operated on a batch basis.

- The Boruah continuous roller. This is one design for rolling continuously and consists of a rotor that turns backwards and forwards through 180° inside a cylinder, grooved internally, which reciprocates along its axis. Leaves fed in one end are moved along and rolled as they travel. Control of the outlet controls the pressure (Boruah, 1975).
- The rotorvane. Devised for continuous rolling by McTear in 1958 (McTear, 1958) this machine is now widely used, frequently in combination with other machines. It consists of a rotor carrying vanes in pairs, each set at right angles to the next. Ribs are provided on the inner surface of the cylinder in which the rotor turns. Leaves fed in at one end are moved along and rolled as they move. Control of the outlet orifice controls the pressure in the machine.
- The cut, tear, curl (CTC) machine. This machine was devised by McKercher *c.* 1932 (Harler, 1964) and has two rollers, carrying teeth, which rotate in opposite directions at different speeds. The teeth mesh in with one another, leaving a gap that is variable. Leaves are drawn through the gap and are cut and curled. It is common practice to reduce the size of the leaves fed to a CTC machine by prior passage through a rotorvane. The cut leaves often pass through a second CTC machine. These machines reduce the leaves to very much smaller pieces than those described earlier. This is important for the fine tea required for tea bags (Fig. 19.8).
- The Lawrie tea processor. This machine operates on a different principle, being essentially a hammer-mill. A rotor carries knives and beaters that cut and beat the leaves on a flat anvil in the casing. Leaves enter the top at one end of the casing and exit through the base at the other end.
- The Legg cutting machine. This machine merely cuts the leaves during their passage under a series of knives. The leaves are not crushed and disrupted sufficiently to produce a tea which is popular in the current market.

FERMENTATION

This starts as soon as the leaves are disrupted. Conversion of flavanols to the coloured theaflavins and thearubigins is an oxidation catalysed by the enzyme polyphenol oxidase. Fermentation therefore requires oxygen from the air. The leaves must also be kept moist during fermentation, and their temperature must be kept low to avoid unwanted substances being formed.

At one time, the rolled leaves were spread on concrete floors to ferment, in rooms into which air, wetted by a water spray, was blown. This system was replaced by a variety of devices (tubs, known in India as 'gumlahs',

Fig. 19.8. A pair of CTC machines. The tea leaves are moved up to the top of the machine by the inclined moving belt. The cutters are behind the tops of the inclined belt guards. The tea falls between the pair of cutters. The first machine on the right discharges cut leaf on to the feed belt of the second machine. (Photograph from George Williamson & Co. Ltd, London.)

continuous belts) which allowed moist air to be blown through a layer of fermenting leaves (Fig. 19.9). The latest development does not involve blowing air through the leaves; a deep bed of leaves moves along and its oxygen content is periodically renewed by dropping the leaves through air.

During fermentation, the tea develops its familiar colour as the theaflavins and thearubigins form. The fermentation must be stopped when these have reached their optimum content and the tea quality has reached its best. This is largely a matter of judgement by experienced tea-makers.

Drying

This operation must first stop the fermentation at its optimum point and then reduce the moisture content of the tea to 2.5–3.5%, at which level the tea will keep its quality when in a sealed container.

Drying is by hot air, which is heated indirectly by a heat exchanger from furnace-flue gases. Fuel for the furnaces is commonly firewood, which is often grown on the plantation; land not required or unsuitable for tea is planted with quick-growing timber species, such as *Eucalyptus* spp.

Early driers involved trays of wet tea, which were static in a hot air current. Continuous driers have been in use for many years. A continuous perforated steel band moves backwards and forwards across the upward

Fig. 19.9. Tea leaves fermenting in mobile tubs (gumlahs). These are moved to connect with a supply of moist, warm air, which blows upwards through the leaves. When fermentation is complete, the tubs are moved to the discharge point, where the tea enters the driers. (Photograph from George Williamson & Co. Ltd, London.)

current of hot air. Wet tea is spread over the band at the top and moves downward, being discharged at the bottom where the air temperature is highest.

The latest development is the fluidized-bed drier. Hot air is blown upwards through a static perforated plate. The velocity of the air is sufficient to lift the tea and move it along the body of the drier. Wet tea introduced at one end leaves the other end dry (Fig. 19.10). A sensor has now been developed which responds to volatile compounds in the drier gas exit. This is connected to the drier controls and regulates conditions to minimize the loss of volatile flavour compounds (Anon., 1998).

SORTING

Tea from the drier must be cleaned to remove foreign matter, stalk and other fibre, and then sorted into the size ranges required for international trade.

Foreign matter is removed by passing the tea over special sieves and over magnets to remove iron. Fibre is removed by passage beneath electrostatically charged rollers, which attract fibre. Finally, a series of sieves separates the tea into the sizes (grades) required for sale. Sometimes, tea is cleaned and sorted by blowing it along a tunnel (the so-called Java tunnel) over bins, in which the different grades collect; the finer the particles, the further they are moved along the row of bins.

Fig. 19.10. Vibro fluid-bed tea drier (photograph from George Williamson & Co. Ltd, London).

PACKING

The traditional package for tea is the plywood tea chest. This rectangular box has a wooden batten on each corner, to which plywood sides, top and base are nailed. This is a very serviceable package, usually made locally in or near the tea-growing area from local timber. It is being replaced, at least for exports to Western countries, by multiwall paper sacks, which are shipped in containers. The change will significantly reduce the pressure for timber from local forests.

McTear (1962) reviewed several improvements in tea manufacture which had been made in the years prior to this review.

Preparation of tea for retail sale

BLENDING

Most of the familiar packeted teas are blends. Skilled tasters select from the available teas that have arrived from producing countries. They aim to produce a consistent appearance and quality throughout the year, although the tea from individual estates inevitably varies with the season of the year. The density of the blends must also suit the standard packages.

Blends for tea bags are made from the very fine grades, 'dusts', which are produced by the modern machines, particularly the CTC machine. Blends are often made up for specific local markets, to suit, for example, the local water-supply.

Retail packaging

This is the only other operation carried out for retail sale. Tea should be handled as little as possible, as movement inevitably causes some breakdown of the particles and makes the tea dusty.

'INSTANT' BEVERAGES

Coffee and tea are now available in immediately soluble or 'instant' form. It takes a measurable time and some care and effort to make an acceptable beverage from the normal products and an insoluble residue remains. A product that dissolves instantly and leaves no residue is therefore extremely attractive. Instant coffee has been widely available for many years, Instant tea has not been available for such a long period.

The preparation of these convenient materials is, at first sight, a simple process. The beverage has first to be made from the normal product and then the water has to be removed to leave a solid residue which can be redissolved when required. In practice, the need for a product which, when redissolved, gives the same taste as the original beverage creates problems. The process must also be efficient in terms of usage of energy and the initial material. The production of instant material therefore becomes a complex chemical-engineering problem. A great deal of effort has gone into developing the details of the processes used and a lot of technology has been patented.

The stages in the production unit are similar for both crops and are as follows.

1. Extraction of soluble materials and removal of insoluble material.
2. Collection of volatile substances which evaporate from the extractor.
3. Concentration of extract liquor.
4. Addition of volatile material to the concentrate and drying to give a stable product.

Each of the two main species of coffee produces a different material when used as the raw material for instant coffee. Arabica coffee produces a better flavour, in line with the product from roasted beans. Robusta coffee, however, produces more soluble solids and so will give more instant material per unit of beans. In the production of instant coffee, therefore, the proportion of each species in the feedstock has to be adjusted to suit the market. The better-flavoured instants will include a higher proportion of arabica and be more expensive; the cheaper products will contain a higher proportion of robusta and be less highly flavoured. The flavour of all instants will be affected by the factors that influence the flavour of beans, particularly the source of beans and the degree of roasting. Instant coffee is now widely used. The taste of some brands may not be the same as coffee direct from beans, but is now accepted in its own right.

Instant tea is not yet widely accepted in the UK and Europe; the taste of quality tea has yet to be equalled. In the USA, it is mainly used as a component in iced drinks. The production of instant tea has been reviewed by Saltmarsh (1991).

REFERENCES

Accorsi, W.R. and Haag, H.P. (1959) Alteracoes morfologicas e citologicas do cafeeiro cultivado em solucao nutritiva, decorrentes des deficiencias e excessas des macronutrientes. *Revista Cafe Portuguesas* 6, 5–19.

Ademosun, O.C. (1993) Performance evaluation of a medium-scale cocoa dehulling and winnowing machine. *Agricultural Mechanization in Asia, Africa and Latin America* 24(2), 57–60, r. 64.

Adeyemi, A.A. (1988) Comparison between some herbicide-mixtures and their pure forms in the control of weeds in old cocoa plantations. In: *Proceedings of Tenth International Cocoa Research Conference, Santo Domingo, Dominican Republic, 1987*, pp. 83–86.

Adeyemi, A.A. (1995) NPK fertilization in rehabilitated cocoa: effects on the growth and development of cocoa under different rehabilitation methods. In: *Summaries of the Twelfth International Cocoa Research Conference, Salvador-Bahia, Brazil, 1995*, p. 70.

Adomako, D. and Tuah, A.K. (1988) Digestibility of cocoa pod husk pectic polysaccharides by sheep fed rations containing cocoa pod husk. In: *Proceedings of the Tenth International Cocoa Research Conference, Santo Domingo, 1987*, pp. 853–858.

Adomako, D., Oppong, H. and Gyedu, E. (1995) Pilot-scale productions of cocoa pod husk ash for the soft soap industry in Ghana. In: *Summaries of the Twelfth International Cocoa Research Conference, Salvador-Bahia, Brazil, 1995*, p. 30.

Adu-Ampomah, Y. (1996) The cocoa breeding programme in Ghana: achievements and prospects for the future. *Cocoa Growers Bulletin* 50, 17–21.

Adu-Ampomah, Y., Novak, F., Afza, R. and van Durren, M. (1990a) Embroid and plant production from cultured cocoa explants. In: *Proceedings of the Tenth International Cocoa Research Conference, Santo Domingo, May, 1987*, pp. 129–136.

Adu-Ampomah, Y., Novak, F., Afza, R. and van Durren, M. (1990b) Determination of methodology to obtain shoot tip culture of cocoa. In: *Proceedings of the Tenth International Cocoa Research Conference, Santo Domingo, May, 1987*, pp. 137–142.

Agnihothrudu, V. (1992) Leaf rust of coffee. In: Muckhopadhyay, A.N. *et al.* (eds) *Plant Diseases of International Importance*, Vol. IV. *Diseases of Sugar, Forest and Plantation Crops*. Prentice-Hall, Englewood Cliffs, New Jersey, pp. 190–201.

Ahenkorah, Y., Halm, B.J., Appiah, M.R., Akrofi, G.S. and Yirenkyi, J.E.K. (1987)

Twenty years' results from a shade and fertilizer trial on Amazon cocoa, *Theobroma cacao* in Ghana. *Experimental Agriculture* 23(1), 31–39.

Alarcon, R. and Carrion, G. (1994) The use of *Verticillium lecanii* as a biological control of coffee rust. *Fitopathologia* 29(1), 82–85.

Alvim, P. de T. (1984) Flowering of cocoa. *Cocoa Growers Bulletin* 35, 28–31.

Alvim, P. de T. (1988) Relacoes entre fatores climaticos e producao do cacaueiro. In: *Proceedings of the Tenth International Cocoa Research Conference, Santo Domingo, Dominican Republic, 1987*, pp. 159–167.

Alvim, P. de T. and Seeschaaf, K.W. (1968) Die-back and death of cocoa plants caused by a new species of parasitic tree. *Nature* 219, 1386–1387.

Alvim, R. (1988) Cacao (*Theobroma cacao* L.) in agrosylvicultural systems. In: *Proceedings of the Tenth International Cocoa Research Conference, Santo Domingo, 1987*, pp. 3–14.

Alvim, R. Pinheiro Lima, J. and Alfonso, C.A. (1982) Viability of bare-root cacao transplanting. In: *Proceedings of the Eighth International Cocoa Research Conference, Cartagena, 1981*, pp. 21–26.

Anandacoomaraswamy, A. and Campbell, G.S. (1993) One dimensional simulation of water use by mature tea. In: *Teatech, 1993, Tea Science and Human Health, Proceedings of the International Symposium, Calcutta*, pp. 219–231.

Anandappa, T.I. (1986) Seed tea. In: Sivapalan, P. and Kathiravetpillai, A. (eds) *Handbook of Tea*. Tea Research Institute of Sri Lanka, Talawakella, Sri Lanka, pp. 10–16.

Ananth, B.R., Iyengar, B.R.V. and Cholakana, N.G. (1965) Widespread zinc deficiency in coffee in India. *Turrialba* 15, 81–87.

Andebrhan, T. (1994) The performance of *Theobroma cacao* trees in Witches Broom disease environments. In: *Proceedings of the Eleventh International Cocoa Research Conference, Yamoussoukro, Côte d'Ivoire, 1993*, pp. 53–56.

Andrade, N.G. (1988a) Fases del Cultivo. In: *Cafetales y Cafe*. Direccion General Sectorial de Information del Sector Agropecuario de Ministero de Agricultura, Caracas, Venezuela, pp. 47–67.

Andrade, N.G. (1988b) Cafetales al sol y la sombra. *Cafetales y Cafe*. Direccion General Sectorial de Information del Sector Agropecuario de Ministero de Agricultura, Caracas, Venezuela, pp. 93–99.

Andrade, N.G. (1988c) Las malezas en los cafetales. *Cafetales y Cafe*. Direccion General Sectorial de Information del Sector Agropecuario de Ministero de Agricultura, Caracas, Venezuela, pp. 103–110.

Andrade, N.G. (1988d) Poda de cafetos. *Cafetales y Cafe*. Direccion General Sectorial de Information del Sector Agropecuario de Ministero de Agricultura, Caracas, Venezuela, pp. 133–140.

Andrade, N.G. (1988e) Enfermedades y plagas del cafeto. *Cafetales y Cafe*. Direccion General Sectorial de Information del Sector Agropecuario de Ministero de Agricultura, Caracas, Venezuela, pp. 143–161.

Aneja, M., Gianfagna, T., Ng, E. and Badilla, I. (1992) Carbon dioxide and temperature influence pollen germination and fruit set in cocoa. *Hortscience* 27(9), 1038–1040.

Aneja, M., Gianfagna, T., Ng, E. and Badilla, I. (1994) Carbon dioxide treatment partially overcomes self-incompatability in a cacao genotype. *Hortscience* 29(1), 15–17.

Anon. (1975) *Diagnostico Socio-economica da Regiao Cacaueira*, Vol. 2. Solos da Regiao Cacaueira.
Anon. (1980) Mistletoes on commercial trees in Colombia. *Haustorium* 6, 3.
Anon. (1983a) *The Pesticide Manual*, 7th Edn (Worthing, C.R. and Walker, S.B., eds). British Crop Protection Society, Croydon, UK.
Anon. (1983b) Mal rosado. In: *Desenvolvimento da Pesquisa e Experimentacao Agropecuaria: Principes Resultado, 1982*. CEPEC, Itabuna, Brazil, pp. 3–5.
Anon. (1993) Agreement on retention plan reached in Kampala. *Newsletter of National Coffee Association of USA* 2430, 23 August 1993.
Anon. (1996) More than a nice cuppa. *Chemistry in Britain* 32(7), 19.
Anon. (1998) Electronic nose helps search for the perfect 'cuppa'. *Developments* 2, 29.
Argunova, V.A. and Pritula, Z.V. (1991) Yield and quality of the tea cultivar Kolkhida as affected by mineral fertilizers. *Vestnik Sel'skokhozyaistvennoi Nauki (Moskva)* 11, 169–172.
Azizah Chulan (Hashim) and Ragu, P. (1986). Growth response of *Theobroma cacao* seedlings to inoculation with vesicular-arbuscular mycorrhizal fungi. *Plant and Soil* 96(2), 279–285.
Balasimha, D. and Rajagopal, V. (1988) Stomatal response of cocoa (*Theobroma cacao*) to climatic factors. *Indian Journal of Agricultural Sciences* 58(3), 213–216.
Banerjee, B. (1987) Can leaf aspect affect herbivory? A case study with tea. *Ecology* 68, 839–843.
Banerjee, M. and Agarwal, B. (1990) *In vitro* rooting of tea, *Camellia sinensis*. *Indian Journal of Experimental Biology* 28(10), 936–939.
Banhammou, D. (1993) La culture du théier au Maroc – développement et perspectives de l'avenir. *Al Awamia* 83, 95–115.
Barbora, A.C., Sarma, J., Barva, D.C. and Barbora, B.C. (1993) Effect of thermal time regulated mechanized harvesting on yield, plucking efficiency and quality of tea harvest. In: *Teatech, 1993. Tea Science and Human Health. Proceedings of the International Symposium*, pp. 270–281.
Barbora, B.C. and Bordoloi, P.K. (1993) Application of remote sensing in soil and water management. In: *Teatech 1993, Tea Science and Human Health, Proceedings of the International Conference, Calcutta*, pp. 232–241.
Barel, M. and Jacquet, M. (1994) Coffee quality: its causes, appreciation and improvement. *Plantations Recherche, Développement* 1(1), 5–13.
Bartley, B.G.D. (1969) Twenty years of cacao breeding at the Imperial College of Tropical Agriculture, Trinidad. In: *Proceedings of the Second International Cocoa Research Conference, Bahia, 1967*, pp. 29–38.
Barua, D.N. (1963) Classification of the tea plant. *Two and a Bud* 10, 3–11.
Barua, D.N. (1987) Tea. In: Sethuraj, M.R. and Raghavendra, A.S. (eds) *Tree Crop Physiology*. Elsevier, Amsterdam, pp. 225–246.
Barua, G.C.S. (1991) Cold weather practices for reducing disease incidence in tea. *Two and a Bud* 38(1–2), 13–17.
Bekele, F.L. (1992) Use of botanical descriptors for cocoa characterization: CRV experiences. In: *Proceedings of International Workshop on Conservation, Characterization and Utilization of Cocoa Genetic Resources in the 21st Century. Trinidad*, pp. 77–103.
Berthouly, M. (1991) Cell suspension and somatic embryogenesis regeneration in liquid medium in *Coffea*. In: *Proceedings of the 14th International Plant Biotechnology Conference, Costa Rica, 1991*.

Berthouly, M., Guzman, N. and Chatelet, P. (1987) Micropropagation *in vitro* of different lines of *Coffea arabica* var. Catimor. In: *Proceedings of the 12th International Coffee Research Conference, Montreux, 1987*, pp. 462–467.

Berthouly, M., Alemanno, L., Michaux-Ferrière, N. and Sartigeo, A. (1995) Somatic embryogenesis of *Theobroma cacao*. In: *Summaries of the Twelfth International Cocoa Research Conference, Salvador-Bahia, Brazil, 1995*, p. 17.

Bezbaruah, B., Bora, T.C. and Saikia, N. (1993) Significance of ureolytic nitrification activity in tea (*Camellia sinensis* (L.)) plantation soils. In: *Teatech 1993. Tea Science and Human Health. Proceedings of the International Symposium*, pp. 293–299.

Bhumibhamon, O., Naka, P. and Julsawad, U. (1993) Cocoa fermentation: study of microbiological, physical and chemical changes during cocoa fermentation. *Kasetsart Journal, Natural Sciences* 27(3), 303–313.

Bieysse, D., Gofflot, A. and Michaux-Ferrière, N. (1993) Effect of experimental conditions and genotypic variability on somatic embryogenesis in *Coffea arabica*. *Canadian Journal of Botany* 71(11), 1496–1502.

Bieysse, D., Bompard, E., Bella Manga, Rousael, V. and Vergnes, C. (1995) Diversité génétique et variabilité du pouvoir pathogène chez *Colletotrichum coffeanum*, agent de l'anthracnose des baies de *Coffea arabica*. In: *Abstracts of the 16th International Coffee Research Conference, Kyoto, 1995*, Abstract AP6.

Blore, T.W.D. (1966) Further studies of water use by irrigated and unirrigated arabica coffee in Kenya. *Journal of Agricultural Science (Cambridge)* 67, 145–154.

Blow, R. (1968) Establishment of cocoa under jungle and conversion to planted shade. *Cocoa Growers Bulletin* 11, 10–12.

Bonaparte, E.E.N.A. (1981) Long-term effects of chemical and manual weed control in cocoa. Parts 1 and 2. In: *Proceedings of the Seventh International Cocoa Research Conference, Douala, 1979*, pp. 91–108.

Boopathy, R. (1989) The overall economics of anaerobic digestion on a coffee estate. 1. Design calculation. *Indian Coffee* 53(3), 3–6.

Bopaiah, B.M., Nagaraja, K.V. and Shantaram, M.V. (1988) Biogas production from cacao pod husk. In: *Proceedings of the Tenth International Cocoa Research Conference, Santo Domingo, 1987*, pp. 857–859.

Boruah, T.C. (1975) The continuous tea roller. In: *Proceedings of the 27th Conference, Tea Research Station, Tocklai Experimental Station, Assam, 13–15 November*, pp. 134–139.

Braham, J.E. and Bressani, R. (1979) *Coffee Pulp; Composition, Technology and Utilisation*. Institute of Nutrition of Central America and Panama.

Brammer, H. (1962) Soils. In: Wills, J.B. (ed.) *Agriculture and Land Use in Ghana*. Oxford University Press, Oxford, pp. 88–126.

Braneau de Mire, P. (1969) An ant used in Cameroon to control the mirid *Wasmania auropunctata* in cocoa. *Café, Cacao, Thé* 13, 209–212.

Brasier, C.M. and Griffin, M.J. (1979) Taxonomy of *Phytophthora palmivora* in cocoa. *Transactions of the British Mycological Society* 72, 111–143.

Briton-Jones, H.R. (1934) *The Diseases and Curing of Cacao*. Macmillan, London.

Browning, G. and Fisher, N.M. (1976) High density coffee; yield results for the first cycle from systematic plant spacing designs. *Kenya Coffee* 41, 209–217.

Bustillo, P.A.E. (1995) Integrated pest management to control the coffee berry borer in

Colombia. In: *Proceedings of the 16th International Coffee Research Conference, Kyoto, 1995*, Abstract A23.

Campos, V.P., Sivapalan, P. and Gnanapragasam, N.C. (1990a) Nematode parasites of coffee. In: Luc, M., Sikora, R.A. and Bridge, J. (eds) *Plant-Parasite Nematodes in Subtropical and Tropical Agriculture*. CAB International, Wallingford, UK, pp. 387–401.

Campos, V.P., Sivapalan, P. and Gnanapragasam, N.C. (1990b) Nematode parasites of cocoa. In: Luc, M., Sikora, R.A. and Bridge, J. (eds) *Plant-Parasite Nematodes in Subtropical and Tropical Agriculture*. CAB International, Wallingford, UK, pp. 401–404.

Campos, V.P., Sivapalan, P. and Gnanapragasam, N.C. (1990c) Nematode parasites of tea. In: Luc, M., Sikora, R.A. and Bridge, J. (eds) *Plant-Parasite Nematodes in Subtropical and Tropical Agriculture*. CAB International, Wallingford, UK, pp. 404–430.

Cannell, M.G.R. (1985a) Role of daylength. In: Clifford, M.N. and Willson, K.C. (eds) *Coffee; Botany, Biochemistry and Production of Beans and Beverage*. Croom Helm, London, p. 110.

Cannell, M.G.R. (1985b) The physiology of the coffee crop. In: Clifford, M.N. and Willson, K.C. (eds) *Coffee; Botany, Biochemistry and Production of Beans and Beverage*. Croom Helm, London, pp. 108–134.

Capot, J. (1972) L'amélioration du caféier, en Côte d'Ivoire. Les hybrides 'arabusta'. *Café, Cacao, Thé* 16, 3–16.

Carr, M.K.V. and Stephens, W. (1991) Climate, weather and the yield of tea. In: Willson, K.C. and Clifford, M.N. (eds) *Tea, Cultivation to Consumption*. Chapman and Hall, London, p. 97.

Carvalho, A., Eskes, A.B., Castillo-Z., J., Sreenivasan, H.S., Echeverri, J.H., Fernandez, C.E. and Fazouli, L.C. (1989) Breeding programs. In: Kushalappa, A.C. and Eskes, A.B.J. (eds) *Coffee Rust; Epidemiology, Resistance and Management*. CRC Press, Boca Raton, Florida.

Carvalho, A., Medina Filho, H.P., Fazouli, L.C., Guerreiro Filho, C. and Lima, M.M.A. (1991) Genetic aspects of the coffee tree. *Revista Brasiliera de Genetica* 14(1), 135–183.

Castillo, M., Bailly, H., Violle, P., Pommares, P. and Salee, B. (1993) Traitement des eaux résiduaires d'usines de transformation du café par voie humide. In: *Proceedings of the 15th Coffee Scientific Conference, Montpellier, 1993*, pp. 370–379.

Castillo, R.E., Arcila, F.J., Jaramillo, R.A. and Sanabria, J. (1995) Light interception in coffee plantations. In: *Proceedings of the 16th International Coffee Research Conference, Kyoto, 1995*, Abstract AP4.

Castro, M. (1990) Progress in research on the coffee borer, *Hypothenemus hampei* in Central America. *Boletin de Promecafe* 49, 7–12.

CBK (1970a) *An Atlas of Coffee Pests and Diseases*. Coffee Board of Kenya, Nairobi, pp. 1–19A, 60–88.

CBK (1970b) *An Atlas of Coffee Pests and Diseases*. Coffee Board of Kenya, Nairobi, pp. 20–34, 100–113.

CBK (1970c) *An Atlas of Coffee Pests and Diseases*. Coffee Board of Kenya, Nairobi, pp. 40–46, 140–146.

CBK (1994) Weed control in coffee. Technical Circular No. 65. *Kenya Coffee* 59(687), 1697–1701.

Chaudhuri, T.C. (1993) Chromosomal complexes in tea (aneuploids and polyploids). In: Mulgy, M.J. and Sharma, V.S. (eds) *Tea – Culture, Processing and Marketing*. Oxford and IBH Publishing, New Delhi, India.

Chenery, E.M. (1955) A preliminary study of aluminium and the tea bush. *Plant and Soil* 6, 174–200.

Chepote, R.E. (1995) The effects of cocoa-pod husk compost on the growth and yield of the cocoa plant. In: *Summaries of the Twelfth International Cocoa Research Conference, Salvador-Bahia, Brazil, 1995*, p. 105.

Chereguino, V.R.S. (1979) Efeto de la poda sanitaria del cafeto y la aplicacion de fungicidas en el control del derrite *Phyllostricta coffeicola*. In: *II Latin American Symposium on Coffee, Garnica, Mexico, 1979*.

Chereguino, V.R.S. (n.d.) Characterization of coffee canker in El Salvador and its possible control. In: *III Simposio Latinoamericano sobre Caficultura*, pp. 224–231.

Chu, H.M. (1975) Studies on soil properties and their grading for tea land improvement. *Taiwan Tea Experiment Station Bulletin* 67, 35–66.

Chude, Y.O. (1988) Comparative performance of five cacao (*Theobroma cacao* L.) cultivars grown on boron deficient subsoils in Nigeria. In: *Proceedings of the Tenth International Cocoa Research Conference, Santo Domingo, Dominican Republic, 1987*, pp. 265–268.

Clapperton, J.F. (1994) A review of research to identify the origins of cocoa flavour characteristics. *Cocoa Growers Bulletin* 48, 7–16.

Clapperton, J.F. (1995) Prospects in the genetics of cocoa flavour. In: *Summaries of the Twelfth International Cocoa Research Conference, Salvador-Bahia, Brazil, 1995*, p. viii.

Clifford, M.N. (1985) Chemical and physical aspects of green coffee and coffee products. In: Clifford, M.N. and Willson, K.C. (eds) *Coffee: Botany, Biochemistry and Production of Beans and Beverage*. Croom Helm, London, pp. 305–374.

Coghlan, A. (1998) Pure brew. *New Scientist* 2126, 11.

Colas, H., Mouchet, S. and Rey, H. (1995) Transpiration of the cacao tree: comparison between isolated cropping and multi-cropping with coconut tree. In: *Summaries of the Twelfth International Cocoa Research Conference, Salvador-Bahia, Brazil, 1995*, p. 77.

Cope, F.W. (1962) The mechanism of pollen incompatibility in *Theobroma cacao*. *Heredity* 17, 157–182.

Cope, F.W. (1976) Cacao. In: Simmonds, N.W. (ed.) *Evolution of Crop Plants*. Longman, London, pp. 285–289.

Cramer, H.H. (1967a) Plant protection and world crop production (translated by J.H. Edwards). *'Bayer' Pflanzenschutz* 20(1), 365.

Cramer, H.H. (1967b) Plant protection and world crop production (translated by J.H. Edwards). *'Bayer' Pflanzenschutz* 20(1), 376.

Cramer, H.H. (1967c) Plant protection and world crop production (translated by J.H. Edwards). *'Bayer' Pflanzenschutz* 20(1), 381.

Cros, E., Bastide, P., n'Guyen Ban, J. and Armengaud, P. (1995) Susceptibility of the cacao tree to mirids: research of biochemical tracers. In: *Summaries of the Twelfth International Cocoa Research Conference, Salvador-Bahia, Brazil, 1995*, p. 10.

Cuenca, G., Herrera, R. and Meneses, E. (1991) Vesicular-arbuscular mycorrhizae and the cultivation of cacao in Venezuela. *Acta Cientifica Venezolana* 42(3), 153–159.

Cunha, J. (1991) The performance of the 'Tubular' dryer, with forced airflow, for cocoa drying. *Agrotropica* 3(1), 39–43.

Cunningham, R.K. and Burridge, J.C. (1960) The growth of cacao (*Theobroma cacao*) with and without shade. *Annals of Botany (New Series)* 24, 458–467.

Daswir, Harris, A.S., and Dja'far (1988) Analysis of cocoa shaded with coconut (*Cocos nucifera*) compared with *Leucaena glauca* in North Sumatra. *Buletin Perkebunan* 19(2), 99–106.

Debry, G. (1994) *Coffee and Health*. John Libbey Eurotext, Paris, 538 pp.

de Castro, F.S. (1960) Relationships between rainfall and coffee production. *Coffee (Turrialba)* 2, 85–89.

de Santana, C.J.L. and Cabala-Rosand, P. (1984) Soil acidity and cocoa response to liming in Southern Bahia, Brazil. *Revista Theobroma* 14(4), 241–251.

Deuss, J. (1969) Influence du mode d'ouverture de plantation, avec ou sans brûlis, sur la fertilité du sol et la productivité des caféiers Robusta en zone forestière Centr-africaine. *Café, Cacao, Thé* 13, 283–289.

Dev Choudhury, M.N. (1993) Effect of phosphorus on the quality of made tea. *Two and a Bud* 40(2), 15–21.

Dey, S.K. (1969) A study of some soils from Assam, Central and East African tea areas. *Two and a Bud* 16(2), 65–69.

Drennan, J.E. and Menzel, C.M. (1994) Synchronization of anthesis and enhancement of vegetative growth in coffee following water stress during floral initiation. *Journal of Horticultural Science* 69(5), 841–849.

Dublin, P. (1980) Inductions de bourgeons néoformes et embryogenèse somatique. Deux voies de multiplication végétative *in vitro* des caféiers. *Café, Cacao, Thé* 24, 121–128.

Dubois, J.C.L. (1987) Impact of agroforestry systems on the integrated development of rural communities in the American tropics. In: *Memorias, Reunion Nacional de Silvicultura, Bogota, Abril 7 al 10 1987*, Abstract 9.

Ducamp, M. (1993) Phytosanitary control of Witches Broom at cacao seed and seedling levels. In: *Summaries of the Twelfth International Cocoa Research Conference, Salvador-Bahia, Brazil, 1995*, p. 14.

Edjamo, Y., Shimber, T., Temesgen, G. and Yilma, A. (1993) Effect of canopies and bearing heads on density and yield of CBD resistant arabica (*Coffea arabica* L.). In: *Proceedings of the 15th International Coffee Research Conference, Montpellier, 1993*, pp. 322–328.

Effendi, S. (1993) Cocoa flavour improvement through pod storage and fermentation using partitioned deep boxes. *Menara Perkebunan* 61(1), 9–12.

Elizondo, J.M. (1988) Evaluation of 12 different types of *Musaceae* as breeding sites for the cacao pollinating insects (*Forcypomia* spp.) in shade and full sun at La Lola, Costa Rica. In: *Proceedings of the Tenth International Cocoa Research Conference, Santo Domingo, 1987*, pp. 297–301.

Entwistle, P.F. (1972) *Pests of Cocoa*. Longman, London, 779 pp.

Erwiyono, R. and Goenadi, D.H. (1990) The potential use of coconut husk material as potting media: growth of cocoa seedlings on coconut husk/sand potting media. *Indonesia Journal of Crop Science* 5(1), 25–34.

Eskes, A.B., Anzveto, F., Peña, M.X. and Anthony, F. (1995) Genetic improvement of coffee for resistance to root-knot nematodes (*Meloidogyne* spp.) in Central America. In: *Proceedings of the 16th International Coffee Research Conference, Kyoto, 1995*, Abstract A19.

Evans, H. (1951) Investigations on the propagation of cacao. *Tropical Agriculture (Trinidad)* 28, 147–203.

Evans, H.C. and Prior, C. (1987) Cocoa pod diseases: causal agents and control. *Outlook on Agriculture* 16(1), 35–41.

Faborode, M.O. (1991) Fermentation and drying of cocoa beans in Nigeria. In: *Proceedings of the International Agricultural Mechanization Conference, Beijing, China, 16–20 October 1991*, pp. 4, 329.

Faborode, M.O. and Dinrifo, R.R. (1994) A mathematical model of cocoa pod deformation based on Hertz theory. *International Agrophysics* 8(3), 403–409.

FAO (1996a) *Yearbook. Production 1996*. FAO, Rome, Table 78, p. 172.

FAO (1996b) *Yearbook. Production 1996*. FAO, Rome, Table 79, p. 173.

FAO (1996c) *Yearbook. Production 1996*. FAO, Rome, Table 80, p. 174.

Fennah, R.G. (1965) Nutritional factors associated with seasonal population increase of cacao thrips (*Selenothrips rubrocinctus*) on cashew. *Bulletin of Entomological Research* 53, 333–349.

Ferwerda, F.P. (1976) Coffees. In: Simmonds, N.W. (ed.) *Evolution of Crop Plants*. Longman, London, pp. 257–260.

Figueira, A. and Janick, J. (1993) Micropropagation of *Theobroma cacao* for germplasm conservation and distribution. In: *Proceedings of International Workshop on Conservation, Characterization and Utilization of Cocoa Genetic Resources in the 21st Century. Trinidad*. Cocoa Research Unit, University of the West Indies, St Augustine, Trinidad, pp. 33–40.

Flood, J. (1997) Coffee under threat. *Spore* 71, 7.

Forestier, A. (1969a) *Culture du Caféier Robusta en Afrique Centrale*. Institut Français du Café et du Cacao, Paris, p. 36.

Forestier, A. (1969b) *Culture du Caféier Robusta en Afrique Centrale*. Institut Français du Café et du Cacao, Paris, p. 37.

Fournier, L.A. (1988) El cultivo del cafeto (*Coffea arabica* L.) al sol o a la sombra: un enfoque agronomico ye ecofisiologico. *Agronomia Costarriconse* 12(1), 131–146.

Fowler, M.S. (1994) Fine or flavour cocoas: current position and prospects. *Cocoa Growers Bulletin* 48, 17–23.

Freeman, W.G. (1929) Cacao research: results of cacao research at River Estate, Trinidad. *Tropical Agriculture* 6, 127–133.

Freeman, W.E. (1969) Some aspects of the cocoa breeding programme. In: *Proceedings of the Agricultural Society of Trinidad and Tobago, December 1968*, pp. 1–15.

Freeman, W.E. (1972) Cocoa breeding. In: *Proceedings of the Fourth International Cocoa Research Conference, Trinidad, 1972*.

Frimpong, E.B., Adu-Ampomah, Y. and Karimu, A.A. (1995) Efforts to breed for drought resistant cocoa in Ghana. In: *Summaries of the Twelfth International Cocoa Research Conference, Salvador-Bahia, Brazil, 1995*, p. 24.

Furlani, A.M., Catani, R.A., de Moraes, F.R.P. and Franco, C.M. (1976) Ecectos de aplicacao de cloreto e de sulfato de potassio na nutricao de cafeeiro. *Bragantia* 35, 349–362.

Gabriel, D., Araujo, J.B.M., Coral, F.J., Tucci, M.L.S., Kupper, R.B., Marints, A.L.M., Saes, L.A. and Júniour, G.G. (1991) Insects associated with pollination of cacao, *Theobroma cacao* L. in Sao Paulo State, Brazil. *Anais da Sociedade Entomologica do Brazil* 20(1), 221–224.

Gabuniya, N.A. (1991) Perspectives of development of machines for tea growing and harvesting. *Traktory i Sel'skokhozyaistvennye Mashiny* 3, 34–38.

Gail Smith, B., Burgess, P.J. and Carr, M.K.V. (1994) Effects of clone and irrigation on the stomatal conductance and photosynthetic rate of tea (*Camellia sinensis*). *Experimental Agriculture* 30, 1–16.

Galindo, J.J. (1992a) Prospects for control of black pod of cacao. In: Keane, P.J. *et al.* (eds) *Cocoa Pest and Disease Management in Southeast Asia and Australasia*. FAO, Rome, Italy.

Galindo, J.J. (1992b) Moniliasis of cocoa in South and Central America. In: Keane, P.J. *et al.* (eds) *Cocoa Pest and Disease Management in Southeast Asia and Australasia*. FAO, Rome, Italy, pp. 37–43.

Galvez, G.C. (1992) Biological control of the Coffee Berry Borer as a viable alternative for Central America. *Boletin de Promocafe* 57, 6–11.

German Medical Information Services (1997) *Chemical and Biological Properties of Tea Infusions* (eds Schubert, R. and Spiro, M.) German Medical Information Services, Frankfurt on Main.

Gething, P.A. (1993) *Improving Returns from Nitrogen Fertiliser: the Potassium–Nitrogen Relationship*. Research Topics, No. 13 (second revision), International Potash Institute, 51 pp.

Glicenstein, L.J., Flynn, W.P. and Fritz, P.J. (1990) Clonal propagation of cocoa. *Cocoa Growers Bulletin* 43, 7–10.

Gonzalez, M.A., Lopez, C.A., Carvapal, J.F. and Briceno, J.A. (1977) Efecto de la fuento de potassio en el acumulamiento de cloruros y sulfatos en el cafeto. *Agronomia Costa Rica* 1, 31–37.

Gopal, N.H., Venkataramanan, D. and Ramaiah, P.K. (1993) Bearing nature of arabica coffee. *Indian Coffee* 57(3), 11–14.

Gouny, P. (1973) Observations on the responses of the plant to the presence of chloride ions. *Potash Review* Subject 3, No. 5.

Govindarajan, T.S. (1988) A review on the incidence of root diseases on coffee and their management. *Journal of Coffee Research* 18(1), (Suppl.), 16–28.

Grandegger, K. (1989) A solar-powered tunnel drier, with collector, for use in coffee, cocoa and coconut production. *Landtechnik* 44(7/8), 293–294.

Green, M.J. (1962) Pre-planting treatment of cuttings. In: *Annual Report, Tea Research Institute of East Africa, 1962*, Kericho, Kenya, pp. 10–11.

Gregory, P.H., Griffin, M.J., Maddison, A.C. and Ward, M.R. (1984) Cocoa blackpod: a reinterpretation. *Cocoa Growers' Bulletin* 35, 5–22.

Grice, W.J. (1984) Mulch – an essential prerequisite for planting. *Quarterly Newsletter, Tea Research Foundation of Central Africa* 75, 16–18.

Haarer, A.E. (1962a) The economic species of coffee. In: *Modern Coffee Production*. Leonard Hill, London, p. 19.

Haarer, A.E. (1962b) Environment and physiology. In: *Modern Coffee Production*. Leonard Hill, London, p. 64.

Hadfield, W. (1968) Leaf temperature, leaf pose and productivity of the tea bush. *Nature* 219, 282–294.

Hadfield, W. (1974a) Shade in North-East Indian tea plantations. *Journal of Applied Ecology* 11, 151–178.

Hadfield, W. (1974b) Foliar illumination and canopy characteristics. *Journal of Applied Ecology* 11, 179–199.

Hanumantha, B.T., Kamanna, B.C., Nirmala Kannan and Govindarajan, T.S. (1994) Effect of chlorothalonil and captafol on brown eye spot disease. *Journal of Coffee Research* 24(1), 15–21.

Harding, P.E. (1993) Seasonal fluctuations in leaf nutrient contents of fertilised and unfertilised arabica coffee in Papua New Guinea. In: *Proceedings of the 15th International Coffee Research Conference, Montpellier 1993*, pp. 799–803.

Harler, C.R. (1964) The rolling of tea leaf. In: *The Culture and Marketing of Tea*, 3rd Edn. Oxford University Press, Oxford, pp. 170–175.

Hasselo, H.N. (1965) The nitrogen, potassium, phosphorus, calcium, magnesium, sodium, manganese, iron, copper, boron, zinc, molybdenum and aluminium contents of tea leaves of increasing age. *Tea Quarterly* 36, 122–136.

Hazarika, U.N., Pradhan, S.K. and Sarkar, S.K. (1965) Effect of extended pruning cycles of varying durations on yield of tea in Darjeeling. *Two and a Bud* 32(1/2), 2–6.

Henderson, T.J. (1966) Hail suppression project utilizing anti-hail rockets. *Tea* 7(1), 11, 13, 15, 17, 19, 21.

Hernandez, E. (1962) El magnesio aumenta los rendimientos del café. *Agricultura al Dia (Puerto Rico)* 36–37.

Herzog, F. (1994) Multiple shade trees in coffee and cocoa plantations in Côte d'Ivoire. *Agroforestry Systems* 27, 259–267.

Hilton, P.J. and Ellis, R.T. (1972) Estimation of market value of Central Africa tea by theaflavin analysis. *Journal of the Science of Food and Agriculture* 23, 227–232.

Holguin, F., Bieysse, D., Eskes, A. and Muller, R. (1993) Virulence and aggressiveness study of *Hemileia vastatrix* isolates from *C. canephora* and Catimor. In: *Proceedings of the 16th International Coffee Research Conference, Montpellier*, pp. 281–292.

Homewood, B. (1995) Cocoa strain breaks fungal spell. *New Scientist* 145(1966), 7.

Huang, T.F. and Chiu, T.F. (1990) Conversion of hand plucking to mechanical harvesting in high grade tea areas in Taiwan. *Acta Horticulturae* 275, 255–260.

Hudson, J.B. (1992) Pruning. *Bulletin – United Planters Association of South India* 45, 29–42.

Hughes, J.d'A. and Ollennu, L.A.A. (1994) Mild strain protection of cocoa in Ghana against cocoa swollen shoot virus. *Plant Pathology* 43(3), 442–457.

IARC–Who (International Agency for Research on Cancer–World Health Organization) (1991a) *Coffee, Tea, Mate, Methylxanthines and Methylglyosal*. IARC Monographs on the Evaluation of Carcinogenic Risks to Humans, Vol. 51, WHO, Rome, p. 295.

IARC–WHO (1991b) *Coffee, Tea, Mate, Methylxanthines and Methylglyosal*. IARC Monographs on the Evaluation of Carcinogenic Risks to Humans, Vol. 51, WHO, Rome, pp. 297–298.

IARC–WHO (1991c) *Coffee, Tea, Mate, Methylxanthines and Methylglyosal*. IARC Monographs on the Evaluation of Carcinogenic Risks to Humans, Vol. 51, WHO, Rome, p. 300.

Idowu, O.L. (1995) Using food crop/cacao intercropping as insect and rodent IPM tool for cocoa production in Nigeria. In: *Summaries of the Twelfth International Cocoa Research Conference, Salvador-Bahia, Brazil, 1995*, p. 4.

International Cocoa Organization (1997) *Quarterly Bulletin of Cocoa Statistics* 23(3), Table 10, pp. 7–8.

International Coffee Organization (1996a) Imports of importing members from all sources. January–December 1991 to 1996. In: *Coffee Statistics, January to December 1987–1992*, ICO, London, Table III(1), p. 29.

International Coffee Organization (1996b) Non-members. Imports of all forms of coffee from all sources. Calendar years 1991 to 1996. In: *Coffee Statistics, January to December 1987–1992*, ICO, London, Table III–IV, p. 39.

International Coffee Organization (1996c) *International Seminar on Coffee and the Environment 27 and 28 May 1986*. Three parts: PR 186/96, EB 3597/96 (E) and Speakers.

International Tea Committee (1997a) Tea retained in producing countries. In: *Annual Bulletin of Statistics, 1996*. ITC, London, p. 48.

International Tea Committee (1997b) Tea imports for consumption. In: *Annual Bulletin of Statistics, 1996*. ITC, London, Table D1, pp. 72–75.

Jadin, P. (1976) Relations entre le potentiel chimique des sols de Côte d'Ivoire et la production des cacaoyers. *Café, Cacao, Thé* 20(4), 287–296.

Jadin, P. (1980) *Mineral Fertilizing – Particularly Potassium – of Cacao in the Ivory Coast Based on Soil Analysis*. International Potash Institute, Berne, Switzerland.

Jadin, P. and Vaast, P. (1990) Estimation of the fertilizer requirements of soils to be used for cocoa plantations in Togo. *Café, Cacao, Thé* 34(3), 179–188.

Jain, S.M., Das, S.C. and Barman, T.S. (1993) Enhancement of root induction from *in vitro* regenerated shoots of tea. *Proceedings of the Indian National Science Academy, Part B, Biological Sciences* 59(6), 623–628.

Jiménez, J.M., Galindo, J.J., Ramírez, C. and Enríquez, G.A. (1988) Evaluation of biological and chemical control to combat *Monilia* Cocoa Pod Rot (*Moniliophthera roreri*) in Costa Rica. In: *Proceedings of Tenth International Cocoa Research Conference, Santo Domingo, 1987*, pp. 453–456.

Jolly, A.L. (1956) Clonal cuttings and seedlings of cocoa. *Tropical Agriculture* 33, 233–237.

Jones, T.A. and Maliphant, G.K. (1958) High yields in cacao field experiments. *Tropical Agriculture (Trinidad)* 35, 272–275.

Joseph, P.A., Potty, N.N., Ashokan, P.K. and Nayar, N.K. (1993) Need for proper management of intercrops in a coconut based cropping system. *Indian Coconut Journal (Cochin)* 24(5), 7–9.

Kabir, S.E., Chaudhuri, T.C. and Hajra, N.G. (1991) Evaluation of herbicides for weed control in Darjeeling tea. *Indian Agriculturist* 35(3), 179–185.

Kahia, N.J. (1993) Plantlet regeneration from *Coffea arabica* shoot tips. *Kenya Coffee* 58(675), 1467–1471.

Kakoty, N.N., Rahman, M.A., Sarmah, M. and Singh, K. (1993) Evaluation of anti-feedant property of a few commercial neem formulations against bunch and psychid caterpillars. *Two and a Bud* 40(1), 18–20.

Karanja, A.M. (1993) Ruiru 11, adoption and performance in the estate sector in Kenya. *Kenya Coffee* 58(679), 1533–1543.

Kasasian, L. and Donelan, A.F. (1965) The effect of herbicides on cocoa (*Theobroma cacao*). *Tropical Agriculture (Trinidad)* 42, 217–222,

Kaufmann, T. (1975) Biology and ecology of *Tyora tessmanni* (Homoptera; Psyllidae) with special reference to its role as cocoa pollinator in Ghana. *Journal of Kansas Entomological Society* 46, 285–293.

Keane, P.J. (1992) Diseases and pests of cocoa: an overview. In: Keane, P.J. and Putter, C.A.J. (eds) *Cocoa Pest and Disease Management in Southeast Asia and Australasia*. FAO, Rome, 1992, pp. 1–11.

Keane, P.J. and Prior, C. (1992) Biology of vascular-streak dieback of cocoa. In: Keane, P.J. and Putter, C.A.J. (eds) *Cocoa Pest and Disease Management in Southeast Asia and*

Australasia. FAO, Rome, 1992, pp. 75–83.
Kebe, B.I. (1988) Study of 'Warty Pod' disease. In: *Proceedings of the Tenth International Cocoa Research Conference, Santo Domingo, Dominican Republic, 1987*, pp. 499–502.
Kennedy, A.J. and Mooleedhar, V. (1992) Conservation of cocoa in field genebanks – the International Cocoa Genebank, Trinidad. In: *Proceedings of International Workshop on Conservation, Characterisation and Utilization of Cocoa Genetic Resources in the 21st Century, Trinidad, 1992*, pp. 21–26.
Kenya Coffee Research Station (1993) Control of coffee berry disease and leaf rust in 1993 (revised). *Kenya Coffee* 5(68), 1652–1658.
Khoo, K.C., Ho, C.T., Ng, K.Y. and Lim, T.K. (1983) Pesticide application technology in perennial crops in Malaysia. In: *Proceedings of Seminar on Pesticide Application Technology, Serdang, 1982*, pp. 42–85.
Kiara, J.M. and Naged, T.F. (1995a) Studies on the management of rust resistant Catimor varieties in Papua New Guinea. In: *Proceedings of the 16th International Coffee Research Conference, Kyoto, 1995*, Abstract AP18.
Kiara, J.M. and Naged, T.F. (1995b) Establishment of rust resistant Arabica coffee cultivar under temporary shade and inorganic fertiliser regimes in Papua New Guinea. In: *Proceedings of the 16th International Coffee Research Conference, Kyoto, 1995*, Abstract AP19.
Kilgour, J. and Burley, J. (1991) Tea mechanisation in Uganda. *Agricultural Engineer* 46(2), 52–55.
Kimemia, J.K. and Njoroge, J.M. (1988) Effect of shade on coffee – a review. *Kenya Coffee* 53(622), 387–391.
Koffi, N. (1995) La nutrition potassique du caféier robusta en Côte d'Ivoire. In: *Proceedings of the 16th International Coffee Research Conference, Kyoto, 1995*, Abstract AP42.
Konam, J.K., Dennis, J., Saul, J., Flood, J. and Guest, D. (1995) Integrated management of *Phytophthora* diseases of cocoa in Papua New Guinea. In: *Summaries of the Twelfth International Cocoa Research Conference, Salvador-Bahia, Brazil, 1995*, p. 68.
Konishi, S. (1990) Stimulatory effects of aluminium on tea plant growth. In: *Transactions of 14th International Congress of Soil Science, Kyoto, Japan, 12–18 August 1990*, Vol. IV, pp. 164–169.
Kramadibrata, K. and Hedger, J.N. (1988) Comparative studies on the mycorrhizal symbionts of cocoa in Ecuador and Indonesia. In: *Proceedings of the Tenth International Cocoa Research Conference, Santo Domingo, 1987*, pp. 509–513.
Kuijt, J. (1964) Critical observations on the parasitism of New World mistletoes. *Canadian Journal of Botany* 42, 1243–1278.
Kulasegaram, S. (1986) Pruning. In: *Handbook on Tea*. Tea Research Institute of Sri Lanka, Talawahele, pp. 82–87.
Kumar, A.C. and Samuel, S.D. (1990) Nematodes attacking coffee and their management – a new review. *Journal of Coffee Research* 10(1), 1–27.
Kumar, V.B.S., Reddy, A.G.S. and Ramachandran, M. (1992) *Parkia roxburghii* – a leguminous shade tree for coffee. *Indian Coffee* 56(4), 3–9.
Kurikanthimath, V.S., Heade, R., Venugopal, M.N., Sivaraman, K. and Krishnamurthy, B. (1994) Multistoreyed cropping systems with coffee, clove (*Syzygium aromaticum*) and pepper (*Piper nigrum*). *Indian Coffee* 58(10), 3–5.
Kushalappa, A.C. and Eskes, A.B.I. (eds) (1989) *Coffee Rust: Epidemiology, Resistance and Management*. CRC Press, Boca Raton, Florida.

Laker, H.A., Sreenivasan, T.N. and Kumar, D.R. (1988) Recent investigations on chemical control of Witches' Broom disease of cocoa in Trinidad. In: *Proceedings of Tenth International Cocoa Research Conference, Santo Domingo, May, 1987*, pp. 331–336.

Lambot, C. and Gahiro, L. (1993) Influence des traitements contre la carence en zinc du caféier arabica sur l'anthracnose des fruits. In: *Proceedings of the 15th International Coffee Research Conference, Montpellier, 1993*, pp. 271–279.

Lanaud, C., Kebe, I., Risterucci, A.M., n'Goran, J.K.A., Grivet, L., Tahi, M., Cilas, C., Pierreti, I., Sokes, A. and Despréaux, D. (1995) The genetic structure of diversity of *Theobroma cacao* and related species revealed by molecular markers and genomic size variations. In: *Summaries of the Twelfth International Cocoa Research Conference, Salvador-Bahia, Brazil, 1995*, p. 19.

Landau, J.M. and Yang, C.S. (1997) The effect of tea on health. *Chemistry and Industry* 17 November, 904–906.

Lass, R.A. (1985) Epiphytic and parasitic plants associated with cocoa. In: Wood, G.A.R. and Lass, R.A. (eds) *Cocoa*. Longman, London, pp. 346–351.

Laup, S. (1994) Pests and diseases of shade trees and their relation to cocoa in Papua New Guinea. In: *Proceedings of the Eleventh International Cocoa Research Conference, Yamoussoukro, Côte d'Ivoire, 1993*, pp. 125–131.

Lavabre, E.M. (1961) *Protection des Cultures de Caféiers, Cacaoyers, et Autres Plantes Pérennes Tropicales*. Institut Français du Café et du Cacao, Paris.

Leite, R.M. de C., Valle, R.R., da Silva, C.P. and Dias, B.R. (1990) Relationships between flowering and fruiting in cocoa. *Agrotropica* 2(1), 11–16.

Le Pelley, R.H. (1968) *Pests of Coffee*. Longmans, Green, London.

le Pierres, D. (1987) Considerations on graft incompatibility in coffee. In: *Proceedings of the Twelfth International Coffee Research Conference, Paris*, pp. 783–790.

le Pierres, D. and Yapo, A. (1993) Système de reproduction des arabusta *lato sensu* et exploitation de leurs descendances pour l'amélioration des caféiers de basse altitude. In: *Proceedings of the 15th International Coffee Research Conference, Montpellier, 1993*, pp. 199–203.

Lim, G.T. (1992) Biology, ecology and control of cocoa podborer *Conopomorpha cramerella*. In: Keane, P.J. and Putter, C.A.J. (eds) *Cocoa Pest and Disease Management in Southeast Asia and Australasia*. FAO, Rome, pp. 85–100.

Lim, G.T. and Chong, T.C. (1986) Biological control of cocoa podborer, *Conopomorpha cramerella*, in Sabah, Malaysia. In: *Proceedings of Second International Conference on Plant Protection in the Tropics (Extended Abstracts)*. Malaysia Plant Protection Society, Kuala Lumpur.

Lim, K.O. and Vizhi, S.M. (1993) Carbonisation of cocoa tree prunings at moderate temperatures. *Bioresource Technology* 44(1), 85–87.

Loué, A. (1957) *La Nutrition Minérale du Caféier en Côte d'Ivoire*. Gouvernement Général de l'AOF, Centre de Recherches Agronomiques, Bingerville, Côte d'Ivoire.

Luis, E. and Sanchez, F. (1991) The weeds of coffee plantations: practical recommendations for their control. *FONAIAP Divulga* 9(38), 18–20.

Luz, E.D.M.N., Ram, A., de Sa, D.F. and Lellis, W.T. (1985) Climatic factors and periodicity of Pink Disease incidence in Bahia, Brazil. In: *Proceedings of the Ninth International Cocoa Research Conference, Lomé, 1984*, pp. 403–409.

McKelvie, A.D. (1956) Cherelle wilt of cacao. 1. Pod development and its relation to

wilt. *Journal of Experimental Botany* 7, 252–263.
McTear, I. (1958) The rotorvane. *Two and a Bud* 5(2), 6–8.
McTear, I. (1962) Recent developments in tea manufacture. *Two and a Bud* 9(2), 4–6.
Maddison, A.C., Rudgard, S.A. and Andebrhan, T. (1994) Summary of research and recommendations from the International Witches Broom Project. In: *Proceedings of the Eleventh International Cocoa Research Conference, Yamoussoukro, Côte d'Ivoire, 1993*, pp. 189–196.
Maduaho, C.N. and Faborode, M.O. (1994) Characterisation of the breaking behaviour of cocoa pods. *Journal of Agricultural Engineering Research* 59(2), 89–96.
Magambo, M.J.S. and Waithaka, K. (1985) The influence of plant density on dry matter production and yield in young clonal tea in Kenya. *Acta Horticulturae* 158, 157–162.
Malavolta, E. (1986) Foliar fertilization in Brazil. In: Alexander, A. (ed.) *Foliar Fertilization*. Developments in Plant and Soil Sciences, Vol. 22, pp. 170–192.
Malavolta, E., Haag, H.P. and Johnson, C.M. (1961) Estudos sobre a alimentacao mineral do cafeiro. VI. Efeitos das deficiencias de micronutrientes en *Coffea arabica* var. *mundo novo* cultivado em solucao nutritiva. *Anais Escola Superior de Agricultura Luis de Queiroz* 18, 147–167.
Malenga, N.E.A. (1996) Response of young unirrigated tea to nitrogen fertilizer. *Experimental Agriculture* 32(1), 63–66.
Malita, J.N. and Mahanta, P.K. (1993) Effect of calcium on boron availability in soil and uptake by young tea. *Two and a Bud* 40(2), 22–25.
Manivel, L. (1978) Importance of maintenance foliage in tea. *Two and a Bud* 25(2), 74–75.
Manivel, L. (1993) Micropropagation of tea clones through tissue culture. *Planters' Chronicle* November, 483–484.
Marks, V. (1991) Physiological and clinical effects of tea. In: Willson, K.C. and Clifford, M.N. (eds) *Tea: Cultivation to Consumption*. Chapman and Hall, London.
Maroko, J.B. (1991) Effects of two annual applications of copper biocides on levels of copper in coffee-plant materials in Bahati-solai, Nakuru, Kenya. *Kenya Coffee* 56(659), 1173–1178.
Marques, D.V. (1993) Induction of somatic embryogenesis on *Coffea eugenioides* by *in vitro* culture of leaf explants. *Café, Cacao, Thé* 37(3), 251–255.
Mbondji, P.M. (1995) Contribution à l'amélioration de la qualité du café par la lutte intégrée contre *Hypothenemus hampei* et *Tetramorium aculeatum*. In: *16th International Coffee Research Conference, Kyoto, 1995*, Abstract A20.
Mehlich, A. (1965) Soil analysis. In: *Annual Report, Coffee Research Foundation, Kenya, 1964/65*, CRF, Thika, Kenya, pp. 46–52.
Mehlich, A. (1967) Mineral nutrition in relation to yield and quality of Kenya coffee. 1. Effect of nitrogen fertilisers, mulch and other materials on yield and grade 'A' quality of coffee. *Kenya Coffee* 32, 399–407.
Melican, J.T.N. (1997) Tea – hand plucking versus mechanical harvesting. *Tropical Agriculture Association Newsletter* 17(4), 2–4.
Mesfin Ameha (1991) Significance of Ethiopian coffee genetic resources to coffee improvement. In: Engels, J. *et al.* (eds) *Plant Genetic Resources of Ethiopia*. Cambridge University Press, Cambridge, UK, pp. 354–359.
Mitchell, H.W. (1974) A study tour of coffee-producing countries. *Kenya Coffee* 39,

24–32, 108–115, 140–147, 164–169, 194–201, 233–237, 283–291.
M'Itungo, A.M. and van der Vossen, H.A.M. (1981) Nutrient requirements of coffee seedlings in polybag nurseries: the effect of foliar feeds in relation to type of potting mixture. *Kenya Coffee* 46, 181–187.
Mohanan, M. and Sharma, V.S. (1981) Morphology and systematics of some tea (*Camellia* spp.) cultivars. In: *Proceedings of the Fourth Symposium on Plantation Crops*, Vol. 4, pp. 391–400.
Mohd Razi, I., Abd Halim, H., Kamariah, D. and Mohd Noh, J. (1992) Growth, plant water relation and photosynthesis of young *Theobroma cacao* as influenced by water stress. *Pertonika Journal of Tropical Agricultural Science* 15(2), 93–97.
Moll, E.R. (1956a) *Instructions for Propagation of Cocoa*. Cocoa Board Circular No. 1 of 1956, Trinidad.
Moll, E.R. (1956b) The pot rooting technique of cacao propagation. In: *Proceedings of 6th Conference Interamericana de Cacao, Salvador, Bahia, 1956*, pp. 221–227.
Montagnon, C. and Leroy, Th. (1993) Résultats récents sur la résistance de *Coffea canephora* à la sécheresse, à la rouille orange et au scolyte des branchettes en Côte d'Ivoire. In: *Proceedings of the 15th International Coffee Research Conference, Montpellier, 1993*, pp. 309–317.
Mooleedhar, V. and Lauckner, F.B. (1990) Effect of spacing on yield in improved clones of *Theobroma cacao*. *Tropical Agriculture, (Trinidad)* 67(4), 376–378.
Morris, J.M. (1997) Estate road systems – principles and pitfalls. *Tropical Agriculture Association Newsletter* 17(4), 13–18.
Motluk, A. (1995) Hill of beans triggers chocolate war. *New Scientist* 13 May, 4.
Mouli, B.C. (1993) *Blister Blight Control – New Recommendations*. UPASI Tea Research Institute, Nirar Dam BPO, Tamel Nadu, India, 8 pp.
Muhammad Nur Ahmad (1992) Cocoa roasting for rural industries. In: *Proceedings of a Conference on Agricultural Engineering and Rural Development, Beijing, 12–14 October, 1992*, Vol. IV, pp. 55–58.
Mukherjee, S. and Singh, K. (1993) Development in tea pest management. *Two and a Bud* 40(1), 2–6.
Muller, M.W., Pinho, A. de F.S. and Alvim, P. de T. (1988) Effect of manual pollination on production and the phenology of cacao. In: *Proceedings of the Tenth International Cocoa Research Conference, Santo Domingo, May 1987*, pp. 275–281.
Muraleedharan, N. (1991) *Pest Management in Tea*. United Planters' Association of Southern India, Tea Research Institute, Nirar Dam BPO, Tamil Nadu, India.
Muraleedharan, N. and Selvasundaram, R. (1991) Bio-ecology of *Phytodietus spinipes* (Cameron) a parasitoid of *Homona coffearia* (Nietner), the tea tortrix. *Journal of Plantation Crops* 19(1), 26–32.
Muraleedharan, N., Radhakrishnan, B. and Selvasundaram, R. (1992) Shot hole borer of tea – evaluation of insecticides. *Planters' Chronicle* February, 79.
Murray, D.B. (1954) A shade and fertilizer experiment with cacao. III. In: *Report of Cacao Research in Trinidad, 1953*, pp. 30–37.
Murthy, S.B.K. (1992) Harvesting in tea – precepts and practices. *Bulletin of United Planters Association of Southern India* 45, 43–62.
Mwakha, E. (1990a) Clonal tea response to enriched filter-press cake: results from potted plants. *Tea* 11(2), 58–64.
Mwakha, E. (1990b) Young clonal tea response to ratios and rates of enriched

manure. *Tea* 11(2), 65–71.

Mwangi, C.N. (ed.) (1983) *Coffee Growers' Handbook*. Coffee Research Foundation, Ruiru, Kenya.

Nakayama, L.H.I. (1989) Effect of boron and zinc rates on the development and mineral nutrition of cocoa trees. *Agrotropica* 1(1), 34–38.

Neog, A.K. (1986) Studies on root growth in tea. PhD thesis, Dibrugarh University, Assam, India.

Ng'etich, W.K. and Othieno, C.O. (1993) Performance of clonal tea plants on rehabilitated moribund tea soils. *Tea* 14(2), 96–106.

Nichols, R. (1964) Studies of fruit development of cacao (*Theobroma cacao*) in relation to cherelle wilt. I. Development of the pericarp. *Annals of Botany (New Series)* 28, 619–635.

Nixon, D.J. and Burgess, P.J. (1996) The effects of mechanical harvesting practices on the yield and leaf quality of clonal tea. In: *First Conference of the Southern and East African Society of Agricultural Engineers, Arusha, Tanzania.*

Njoroge, J.M. (1989) A review of some agronomic investigations in Kenya. *Kenya Coffee* 54(629), 553–567.

Njoroge, J.M., Waithaka, K. and Cheveya, J.A. (1993) Effects of intercropping young plants of the cultivar Ruiru 11 with potatoes, tomatoes, beans and maize on coffee yields and economic returns in Kenya. *Experimental Agriculture* 29(3), 373–377.

Njuguna, C.K. (1993) Suitability of some Kenyan tea clones for replanting and effect of soil rehabilitation on replanted tea. *Tea* 14(1), 7–12.

Njuguna, C.K. and Magambo, M.J.S. (1993) Effect of rejuvenation pruning and increasing plant density on yields of old tea bushes interplanted with clonal tea plants. *Tea* 14(1), 13–20.

Nutman, E.J. (1933) The root system of *Coffea arabica* Part I. Root systems in tropical soils in British East Africa. *Empire Journal of Experimental Agriculture* 1, 271–284.

Nyirenda, H.E. (1994) The effect of splitting nitrogen fertilizer applications on tea quality and crop distribution of clonal tea under rain-fed conditions. *Quarterly Newsletter, Tea Research Foundation of Central Africa* 115, 3–6.

O'Donohue, J.B. (1992) Practical integrated control of pests and diseases of cocoa in Papua New Guinea. In: Keane, P.J. and Putter, C.A.J. (eds) *Cocoa Pest and Disease Management in Southeast Asia and Australasia*. FAO, Rome, pp. 69–74.

Oduwole, O.O. and Arueya, G.I. (1990) Potential for potash production from cocoa-pod husk in Nigeria. *Agrotropica* 22(3), 171–175.

Okali, D.U.U. and Owusu, J.K. (1975) Growth analysis and photosynthetic rates of cocoa (*Theobroma cacao* L.) seedlings in relation to varying shade and nutrient regimes. *Ghana Journal of Agricultural Science* 8, 51–67.

Ollennu, L.A. and Owusu, G.K. (1988) Studies of the reinfection of replanted cocoa by cocoa swollen shoot virus in Ghana. In: *Proceedings of the Tenth International Cocoa Research Conference, Santo Domingo, 1987*, pp. 515–520.

Omondi, C.O. and Owuor, J.B.O. (1992) The performance of interspecific crosses and backcrosses involving arabica and tetraploid robusta species. *Kenya Coffee* 57(666), 1307–1312.

Onsando, J.M. and Waudo, S.W. (1994) Interaction between *Trichoderma* isolates and *Armillaria* root rot fungus of tea in Kenya. *International Journal of Pest Management* 40(1), 69–74.

Ooi, P.A.C. (1992) Prospects for biological control of cocoa insect pests. In: Keane, P.J. and Putter, C.A.J. (eds) *Cocoa Pest and Disease Management in Southeast Asia and Australasia*. FAO, Rome.

O'Shea, P.B.T. (1964) Placement of fertilisers. In: *Annual Report 1963*. Tea Research Institute of East Africa, Kericho, Kenya, p. 27.

Othieno, C.O. (1979) Soil temperature and the growth of the tea plant. *Tea in East Africa* 18(1), 5–7.

Othieno, C.O. (1982) Supplementary irrigation of young clonal tea in Kenya. III. Comparative dry matter production and partition. *Tea* 3, 15–25.

Othieno, C.O. (1991a) Mean annual rainfall and lowest and highest mean monthly temperatures in tropical and sub-tropical tea growing areas. In: Willson, K.C. and Clifford, M.N. (eds) *Tea: Cultivation to Consumption*. Chapman and Hall, London, p. 138.

Othieno, C.O. (1991b) General chemical properties of some tea soils. In: Willson, K.C. and Clifford, M.N. (eds) *Tea: Cultivation to Consumption*. Chapman and Hall, London, p. 139.

Othieno, C.O. (1991c) Physical properties of some tea soils of the world. In: Willson, K.C. and Clifford, M.N. (eds) *Tea: Cultivation to Consumption*. Chapman and Hall, London, p. 156.

Othieno, C.O. and Ng'etich, W.K. (1993) Studies on the use of shade in tea plantations in Kenya. II Effect on yields and components. In: *Teatech, 1993. Tea Science and Human Health. Proceedings of the International Symposium, Calcutta*, pp. 282–292.

Owuor, P.O. (1993) Comparative studies of black tea quality and chemical composition of different varieties and clones of *Camellia sinensis* in Kenya – a review. *Tea* 14(1), 54–60.

Owuor, P.O., Gone, F.C., Onchiri, D.B. and Jumba, I.C. (1990) Levels of aluminium in green leaf of clonal teas, black tea and black tea liquors, and effects of rates of nitrogen fertilisers on the aluminium black tea contents. *Food Chemistry* 35(1), 59–68.

Owusu, G.K., Ollennu, L.A.A. and Dzahini-Obatey, M. (1995) The prospects for mild-strain cross-protection to control cocoa swollen-shoot disease in Ghana. In: *Summaries of Twelfth International Cocoa Research Conference, Salvador-Bahia, Brazil, 1995*, p. 29.

Panton, W.P. (1957) Soil Survey Reports, No. 5. Federal Experiment Station, Jerangau, Trengganu. *Malayan Agricultural Journal* 40, 19–29.

Pavan, M.A., Bingham, F.T. and Pratt, P.F. (1982) Toxicity of aluminium to coffee in ultisols and oxisols amended with $CaCO_3$, $MgCO_3$ and $CaSO_4.2H_2O$. *Soil Science Society of America* 46, 1201–1206.

Penman, H.L. (1948) Natural evaporation from open water, soil and grass. *Proceedings of the Royal Society, Series A* 193, 120–145.

Peregrine, W.T.H. (1991) Annatto – a possible trap crop to assist control of the mosquito bug (*Helopeltis schoutedenii*) in tea and other crops. *Tropical Pest Management* 37(4), 429–430.

Persad, C. (1988) Studies on the control of black pod disease of cocoa in Trinidad. In: *Proceedings of the Tenth International Cocoa Research Conference, Santo Domingo*, pp. 437–440.

Pinto, L.R.M., Pereira, M.G., Carletto, G.A. and dos Santos, A.V.P. (1990) Genetic improvement of cocoa: methods for acquisition and use of endogamous pro-

genitors for hybrid production. *Agrotropica* 2(2), 59–67.
Ponnapa, P.S. (1991) Replanting – practices and economics. *Bulletin – United Planters Association of Southern India* 44, 77–84.
Posnette, A.F. (1940) Transmission of swollen shoot. *Tropical Agriculture (Trinidad)* 17, 98.
Rahman, F. (1991) Effect of different forms of cultivation and chemical weed control on yield of tea. *Two and a Bud* 38(1–2), 28–30.
Rahman, F., Fareed, M. and Saikia, P. (1981) Effect of bush population on yield of tea. *Journal of Plantation Crops* 9(2), 100–104.
Rakotomalala, J.-J.R., Cros, E., Charrier, A., and Anthony, F. (1993) Marqueurs biochimiques de la diversité des caféiers. In: *Proceedings of the 15th International Coffee Research Conference, Montpellier, 1993*, pp. 47–53.
Ramachandran, M., Reddy, A.G.S. and Kumar, V.B.S. (1993a) 'Top working' of *C. canophora* 'Robusta' cv. 5274 with *C. arabica* cv. 'Cauvery'. *Journal of Coffee Research* 23(2), 121–125.
Ramachandran, M., Kumar, V.B.S. and Reddy, A.G.S. (1993b) Rejuvenation of old coffee with newer selections by cleft grafting – a preliminary report. *India Coffee* 57(4), 24–28.
Rao, W.K. and Ramaiah, P.K. (1986) An approach to rationalised fertiliser usage for coffee. In: *Proceedings of the Eleventh International Coffee Research Conference, Paris*, pp. 589–598.
Redshaw, M.J. and Zulnerlin, Ir., (1996) Cocoa cultivation on PTPP London Sumatra Indonesia estates in North Sumatra province, Indonesia. *Cocoa Growers' Bulletin* 49, 7–25.
Robert, (1990) *Les Vertus Thérapeutiques du Chocolat*. Editions Artulen, Paris, France.
Robertson, A. (1991) The chemistry and biochemistry of black tea production. In: Willson, K.C. and Clifford, M.N. (eds) *Tea, Cultivation to Consumption*. Chapman and Hall, London, p. 556.
Robinson, J.B.D. (1960) Amber beans. *Kenya Coffee* 25, 91–95.
Robinson, J.B.D. (1964) *A Handbook on Arabica Coffee in Tanganyika*. Tanganyika Coffee Board, Moshi, Tanganyika, pp. 22–23.
Robinson, J.B.D. and Hosegood, P.A. (1963) Effects of organic mulch on fertility of a latosolic coffee soil in Kenya. *Experimental Agriculture* 1, 67–80.
Rodrigues Filho, Carrarão, A.P., Lourenço and Júnior, J. de B. (1993) *Evaluation of Agroindustrial Byproducts for Feeding Ruminants*. Documentos No. 71, Centro de Pesquisa Agropecuaria do Tropico Umido, 15 pp.
Room, P. (1972) Mistletoe on West Africa cocoa. *Cocoa Growers' Bulletin* 18, 14–18.
Rosand, P.C. (1995) The nutritional requirements of cocoa under different production systems. In: *Summaries of the Twelfth International Cocoa Research Conference, Salvador-Bahia, Brazil, 1995*.
Rudgard, S.A., Maddison, A.C. and Andebrhan, T. (1993) *Disease Management in Cocoa: Comparative Epidemiology of Witches' Broom*. Chapman and Hall, London.
Ruinard, J. (1966) Notes on the importance of proper husbandry in some tropical crops. *Netherlands Journal of Agricultural Science* 14, 263–279.
Saleh, M. (1973) The relationship between age and mineral content of *Theobroma cacao* leaves. *Menara Perkebunan* 41(5), 219–222.
Saltmarsh, M. (1991) Instant tea. In: Willson, K.C. and Clifford, M.N. (eds) *Tea: Cultivation to Consumption*. Chapman and Hall, London, pp. 534–554.

Samah, O.A., Nor, O.M., Rahman, A.R.A. and Saleh, A.B. (1992) Cocoa, pineapple and sugar cane waste for ethanol production. *Planter (Malaysia)* 68(792), 125–128.

Samuel, S.D., Balakrishnan, M.M. and Bhat, P.K. (1993) A review on brown scale (*Saissetia coffeae*) in India. *Indian Coffee* 57(1–2), 23–24.

Sandaman, S.S., Krishnapillai, S. and Sabaratnam, J. (1978) Nitrification of ammonium sulphate and urea in an acid yellow podzolic tea soil in Sri Lanka in relation to soil fertility. *Plant and Soil* 49, 9–22.

Sanderson, G.W. (1963) The chloroform test – a study of its suitability as a means of rapidly evaluating fermenting properties of cones. *Tea Quarterly* 134, 173–176.

Sarathchandra, T.M., Upali, P.D. and Arulpragasam, P.V. (1990) Progress towards the commercial propagation of tea by tissue culture techniques. *Sri Lanka Journal of Tea Science* 59(2), 62–64.

Satyanarayana, N. (1991) Response to two-stage tipping of young tea in Nilgiris. *Bulletin of United Planters Association of Southern India* 44, 61–68.

Satyanarayana, N. and Sharma, V.S. (1986) Tea (*Camellia* spp.) germplasm in South India. In: Srivastava, H.C., Vatgya, B. and Menon, K.K.G. (eds) *Plantation Crops: Opportunities and Constraints.* Oxford and IBH Publishing, New Delhi, India, pp. 173–179.

Sena Gomes, A.R., Batista, M.L.R., Campelo, C.J.E. and Afonso, C.A. (1995) Analysis of the pruning of cocoa-plant hybrids to different height. In: *Summaries of the Twelfth International Cocoa Research Conference, Salvador-Bahia, Brazil, 1995*, p. 115.

Senaratne, K.A.D.W. (1986) Insect and mite pests of tea. In: Sivapalan, P., Kulasegaram, S. and Kathiravetpillai, A. (eds) *Handbook on Tea.* Tea Research Institute of Sri Lanka, Talawakelle, pp. 88–108.

Sha, J.Q. and Zheng, D.X. (1994) Study of the fluorine content in fresh leaves of tea plant in Fujian province. *Journal of Tea Science* 14(1), 37–42.

Sharma, V.S. and Satyanarayana, N. (1993) Harvesting in tea. In: *Teatech 1993, Tea Science and Human Health, Proceedings of the International Symposium, Calcutta, 1993*, pp. 242–258.

Shuvalov, Yu.N. and Konidze, A.N. (1987) Uptake of selenium by tea plants from virgin and fertilized red soils. *Subtropicheske Kul'tury* 6, 52–54.

Shuvalov, Yu.N. and Sardzhveladze, A.G. (1990) The effect of potassium fertilization on Ca uptake by tea plants. *Subtropicheskie Kul'tury* 2, 47–50.

Silva-Acuna, R., Zambolïm, L. and Gonzalez-Molina, E. del C. (1993) Control of coffee leaf rust under shade by triadimenal combined with copper oxychloride in Venezuela. *Summa Phytopathologica* 19(3–4), 189–194.

Simmonds, N.W. (1993) The breeding of perennial crops. In: *Proceedings of International Workshop on Conservation, Characterization and Utilization of Cocoa Genetic Resources in the 21st Century, Trinidad, 1992.* Cocoa Research Unit, University of the West Indies, St Augustine, Trinidad, pp. 156–162.

Simmonds, N.W. (1994) Horizontal resistance to cocoa diseases. *Cocoa Growers' Bulletin* 47, 42–52.

Singh, B. (1978) Prediction of drainage and irrigation requirements of tea based on meteorological data. *Two and a Bud* 25(2), 75–79.

Singh, I.D. (1992) Release of biclonal seed stock TS 520. *Two and a Bud* 39(1), 46.

Singh, K. (1992a) Clonal susceptibility of red rust. *Two and a Bud* 39(2), 52.

Singh, K. (1992b) Revised list of approved pesticides (as on December 1992). *Two and a Bud* 39(2), 56–62.

Siqueira, P.R., Augusto, S.B., Dias, L.A.S. and Sowza, C.A.S. (1995) The effects of irrigation on the productivity of the cacao plant in Linhares-Es, Brazil. In: *Summaries of the Twelfth International Cocoa Research Conference, Salvador-Bahia, Brazil, 1995*, p. 118.

Sitapai, E.C., Kennedy, A.J. and Kumar, D.R. (1988) Genetic aspects of resistance to *Phytophthora palmivora* disease. In: *Proceedings of the Tenth International Cocoa Research Conference, Santo Domingo, 1987*, pp. 633–641.

Sivapragasam, A., Musa, M.J. and Azmi, M. (1992) Incidence of insect pests in stored cocoa beans and their control using methyl bromide. In: Ooi, P.A.C. *et al.* (eds) *Proceedings of the Third International Conference on Plant Protection in the Tropics.* pp. 64–68.

Smale, P.E., (1991) New Zealand has its own green tea industry. *Horticulture in New Zealand* 2(2), 6–9.

Small, L.E. and Horrell, R.S. (1993) High yield coffee technology. In: *Proceedings of the 15th International Coffee Research Conference, Montpellier, 1993*, pp. 719–725.

Smith, A.N. (1960) Boron deficiency in *Grevillea robusta. Nature* 186(4729), 987.

Smith, E.S.C. (1981) Review of control measures for *Pantorhytes* (Coleoptera: Curculionidae) in cocoa. *Protection Ecology* 3, 279–297.

Smith, E.S.C. (1985) A review of relationships between shade types and cocoa pest and disease problems in Papua New Guinea. *Papua New Guinea Journal of Agriculture, Forestry and Fisheries* 33(3/4), 79–88.

Smith, R.I., Harvey, F.J. and Cannell, M.G.R. (1990) Pattern of tea shoot growth. *Experimental Agriculture* 26(2), 197–208.

Smyth, A.J. and Montgomery, R.F. (1962). *Soils and Land Use in Central Western Nigeria.* Government of Western Nigeria, Ibadan.

Snoeck, J. (1963) La taille du caféier robusta à Madagascar. *Café, Cacao, Thé* 7, 421–432.

Snoeck, J. (1978) Utilisation of diuron in coffee and cocoa nurseries in Côte d'Ivoire. In: *Proceedings of the Third Symposium on Weeding in Tropical Crops, Dakar*, Vol. II, pp. 340–346.

Snoeck, J. (1984a) Coffee. In: Martin-Prével, P. *et al.* (eds) *Plant Analysis as a Guide to the Nutrient Requirements of Temperate and Tropical Crops.* Lavoisier, Paris, pp. 418–431.

Snoeck, J. (1984b) Cacao. In: Martin-Prével, P. *et al.* (eds) *Plant Analysis as a Guide to the Nutrient Requirements of Temperate and Tropical Crops.* Lavoisier, Paris, pp. 432–439.

Snoeck, D., Bitoga, J.P. and Barantwaririje, C. (1994) Advantages and drawbacks of different types of soil cover in coffee plantations in Burundi. *Café, Cacao, Thé* 38(1), 41–48.

Somarriba, E. (1992) Timber harvest damage to crop plants, and yield reduction in two Costa Rican coffee plantations with *Cordia alliodora* shade trees. *Agroforestry Systems* 18(1), 69–82.

Söndahl, M.R., Liu, S., Bellato, C. and Bragin, A. (1993) Cacao somatic embryogenesis. *Acta Horticulturae* 336, 245–248.

Soria, S. de J., Pinho, A.F. de S. and Peixoto, E.S. (1986) Cocoa pollination in the Salvador Bay region of Bahia, Brazil. 2. Assessment of the forced ventilation method. *Revista Theobroma* 16(1), 31–38.

Sosamma, V.K., Koshy, P.K. and Sandararaju, P. (1980) Plant parasitic nematodes associated with cocoa. *Cocoa Growers' Bulletin* 29, 27–30.

Squire, G.R. (1979) Weather, physiology and seasonality of tea (*Camellia sinensis*) yields in Malawi. *Experimental Agriculture* 15, 321–330.
Squire, G.R. and Callander, B.A. (1981) Tea plantations. In: Koslowski, T.T. (ed.) *Water Deficits and Plant Growth*, Vol. 6. Academic Press, New York, pp. 471–510.
Sreenivasan, T.N. (1995) Grafting on very young cacao seedlings. In: *Summaries of the Twelfth International Cocoa Research Conference, Salvador-Bahia, Brazil, 1995*, p. 82.
Sreenivasaprasad, S., Brown, A.E. and Mills, P.R. (1993) Coffee berry disease pathogen in Africa: genetic structure and relationship to the group species *Colletotrichum gloeospioides*. *Mycological Research* 97(8), 995–1000.
Starisky, G. (1970) Embryoid formation in callus tissues of coffee. *Acta Botanica* 19(4), 509–514.
Stensvold, I., Tverdal, A. and Jacobsen, B.K. (1996) Cohort study of coffee intake and death from coronary heart disease. *British Medical Journal* 312, 544–545.
Stephens, W., Burgess, R.J. and Carr, M.K.V. (1994) Yield and water use of tea in Southern Tanzania. *Aspects of Applied Biology* 38, 223–230.
Sudoi, V. (1990) Relative resistance-susceptibility of some Kenyan tea clones to red spider mite *Olygonychus coffeae* – preliminary indications. *Tea* 11(1), 25–28.
Sugha, S.K., Singh, B.M. and Sharma, K.L. (1990) Factors affecting the development of blister blight of tea. *Indian Phytopathology* 43(2), 226.
Sugha, S.K., Singh, B.M., Sharma, D.K. and Sharma, K.L. (1991) Effect of blister blight on tea quality. *Journal of Plantation Crops* 19(1), 59–60.
Sugiyono (1989) Critical level of aluminium in the soil for cocoa. *Buletin Perkebunan* 20(4y), 157, 175–186.
Sugiyono (1992) Status of K nutrient in North Sumatra. *Buletin Perkebunan* 23(2), 64, 91–102.
Sumbak, J.M. (1983) Sulfur requirements of tropical tree crops. In: Blair, G.J. and Till, A.R. (eds) *Sulfur in South-east Asian and South Pacific Agriculture*. University of New England, Armidale, pp. 65–75.
Swaminathan (1992) Integrated nutrient management in tea. *Bulletin of United Planters Association of Southern India* 45, 13–28.
Sylvain, P.G. (1959) El cafeto en relacion el aqua. *Materiales de Ensenanza de Cafe y Cacao* 11, 1–46.
Tabageri, L.G., Mikeladze, A.D. and Cheishvili, I.D. (1991) Results of studies on measures for rehabilitating depressed tea plantations on podzolic soil. *Subtropicheskie Kul'tury* 4, 136–141.
Tahardi, J.S. (1994) Development of *in vitro* techniques for clonal propagation of estate crops. *Buletin Bioteknologi Perkebunan* 1(1), 3–9.
Tan, C.H., Chan, C.L. and Tay, S.P. (1991) Commercial establishment of papaya intercropping with cocoa – Asiatic experience. *Planter* 67(784), 301–313.
Tanton, T.W. (1982a) Environmental factors affecting the yield of tea (*Camellia sinensis*). *Experimental Agriculture* 18, 47–52.
Tanton, T.W. (1982b) Environmental factors affecting the yield of tea (*Camellia sinensis*). II Effects of soil temperature, day length and dry air. *Experimental Agriculture* 18, 53–63.
Tanton, T.W. (1991) Effect of day length on shoot growth. In: Willson, K.C. and Clifford, M.N. (eds) *Tea: Cultivation to Consumption*. Chapman and Hall, London, pp. 193–196.
Tavartkiladze, O.K. (1990) Micropropagation of tea *in vitro*. *Subtropichiskie Kul'tury* 1, 40–47.

Tay, E.B., Lee, M.T., Bang, C.L., Chong, T.C., Lee, Y.M., Lo, D.S.F., Tulas, M.T., Phua, P.K. and Simin, D. (1989) Integrated management of VSD in mature cocoa. *Technical Bulletin, Department of Agriculture, Sabah* 9, 57–70.

Tea Research Association, India (1993) Country profiles. In: *Teatech; Proceedings of the International Symposium on Tea Science and Human Health, Calcutta, January 1993*, pp. 7–106.

Tea Research Institute of East Africa (1968) Summary: bringing tea plants into bearing. In: *TRIEA Annual Report, 1966/67*. Tea Boards of Kenya, Uganda and Tanzania, Kericho, Kenya, pp. 68–69.

Teisson, C. (1994) *La Culture* in Vitro *de Plantes Tropicales*. CIRAD-GERDAT, Montpellier.

Thorald, C.A. (1953) Observations on fungicide control of witches' broom, black pod and pink diseases of *Theolorama caeao*. *Annals of Applied Biology* 40, 362–376.

Thorold, C.A. (1975) Cushion gall diseases. In: *Diseases of Cocoa*. Clarendon Press, Oxford, pp. 134–143.

Toruan-Mathis, N. and Sumaryono (1994) Application of synthetic seed technology for mass clonal propagation of crop plants. *Buletin Bioteknologi Perkebunan* 1, 10–16.

Toxopeus, H. (1972) Cocoa breeding: a consequence of mating system, heterosis and population structure. In: *Cocoa and Coconuts in Malaysia. Proceedings of Conference of Incorporated Society of Planters, Kuala Lumpur, 1971*, pp. 3–12.

TRFK (Tea Research Foundation of Kenya) (1986a) Windbreaks. In: *Tea Growers' Handbook*, 4th edn. TRFK, Kericho, Kenya, pp. 27–29.

TRFK (1986b) Plucking, pruning, skiffing and tipping in mature tea. *Tea Growers' Handbook*, 4th edn. TRFK, Kericho, Kenya, pp. 81–88.

TRFK (1986c) Fertilizers for mother bushes. *Tea Growers' Handbook*, 4th edn. TRFK, Kericho, Kenya, pp. 120–121.

TRFK (1986d) Fertilizer placement in planting holes. *Tea Growers' Handbook*, 4th edn. TRFK, Kericho, Kenya, pp. 123–125.

TRFK (1986e) Fertilizers for seed bearers. *Tea Growers' Handbook*, 4th edn. TRFK, Kericho, Kenya, pp. 133–136.

TRFK (1986f) Treatment of hutsites and soils of pH higher than optimum. *Tea Growers' Handbook*, 4th edn. TRFK, Kericho, Kenya, pp. 136–138.

TRFK(1986g) Diseases, pests, weed control and other abnormalities. *Tea Growers' Handbook*, 4th edn. TRFK, Kericho, Kenya, pp. 163–189.

TRFK (1986h) Treatment of high pH soil for nursery use. *Tea Growers' Handbook*, 4th edn. TRFK, Kericho, Kenya, p. 139.

TRFK (1986i) Sulphur. *Tea Growers' Handbook*, 4th edn. TRFK, Kericho, Kenya, pp. 137–138.

Triana, J.V. (1957) Informe preliminar sobre un estudio de 'modalides del cultivo del cafeto. *Cenicafé* 6, 156–168.

Tsuji, M., Kuboi, T. and Konishi, S. (1994) Stimulating effects of aluminium on cultured roots of tea. *Soil Science and Plant Nutrition* 40(3), 471–476.

Unlu, M.Y., Topcuoglu, S., Kucukcessar, R., Varinlioglu, A., Gungor, N., Bulut, A.M. and Gungot, E. (1995) Natural effective life of ^{137}Cs in tea plants. *Health Physics* 68(1), 94–99.

Urgert, R., Meyboom, S., Kuilman, M., Rexwinkel, H., Vissers, M.N., Klerk, M. and Katan, M.B. (1996) Comparison of effect of cafetiere and filtered coffee on serum concentrations of liver aminotransferases and lipids: six month randomised controlled trial. *British Medical Journal* 313, 1362–1366.

Vaast, P. (1990) Presentation of cacao regeneration systems on smallholdings in Togo. In: *Proceedings of the Tenth International Cocoa Research Conference, Santo Domingo*, pp. 63–69.

van Boxtel, J.H.J. (1994) Studies on genetic transformation of coffee by using electroporation and the biolistic method. Thesis, Wageningen, The Netherlands.

van der Graaf, N.A. (1992) Coffee berry disease. In: Muckhopadhyay, A.N. *et al.* (eds) *Plant Diseases of International Importance*, Vol. 4. Prentice-Hall, Englewood Cliffs, New Jersey.

van der Vossen, H.A.M. (1979) Methods of preserving the viability of coffee seed in storage. *Kenya Coffee* 45, 31–35.

van der Vossen, H.A.M. (1985) Coffee selection and breeding. In: Clifford, M.N. and Willson, K.C. (eds) *Coffee: Botany, Biochemistry and Production*. Croom Helm, London, pp. 48–96.

van Hall, C.J.J. (1914) *Cocoa*. Macmillan Press, London.

Verma, D.P. and Sharma, V.S. (1995) Cost-effective nutrition for sustained tea production in South India. *Fertilizer News* 40(2), 11–19.

Viroux, R. and Petithuguenin, P. (1993) Capping of robusta coffee bushes: an economically advantageous practice. *Café, Cacao, Thé* 37(1), 21–34.

Waikwa, J.W. and Mathenge, W.M. (1977) Field studies on the effect of *Bacillus thuringiensis* on the larvae of the giant coffee looper, *Ascotis scelenario reciprocaria*, and its side effects on the larval parasites of the leaf miner (*Leucoptera* spp.). *Kenya Coffee* 42, 95–101.

Waiyaro, D.J.A. (1983) Considerations in breeding for improved yield and quality in arabica coffee. Doctoral thesis, Wageningen, the Netherlands, p. 9.

Waller, J.M. (1988) Coffee diseases: current status and recent developments. In: Raychaudhuri, S.P. and Verma, J.P.I. (eds) *Review of Tropical Plant Pathology*. International Mycological Institute, Egham, pp. 1–33.

Waller, J.M. and Holderness, M. (1997) Beverage crops and palms. In: Hillocks, R.J. and Waller, J.M. (eds) *Soilborne Diseases of Tropical Crops*. pp. 229–253.

Waller, J.M., Bridge, P.D., Black, R. and Haliza, G. (1993) Characterization of the coffee berry disease pathogen, *Colletotrichum kahawae* sp. nov. *Mycological Research* 97(8), 989–994.

Wallis, J.A.N. (1963) Water use by irrigated coffee in Kenya. *Agricultural Science* 60, 381–388.

Wamatu, J.N. (1990) Vegetative propagation of Ruiru 11: a review. *Kenya Coffee* 55(650), 983–985.

Wang, J.K. and Shellenberger, F.A. (1967) Effect of cumulative damage due to stress cycles in selective harvesting of coffee. *Transactions of the American Society of Agricultural Engineers* 10, 252–255.

Wardojo, S. (1984) Kemungkinan pembebasan Maluku Utara daripada Masalah penggerek buah cokelat *Acrocercops cramerella*. *Menara Perkebunan* 57, 77–83.

Warren, J. (1992) Cocoa breeding in the 21st century. In: *Proceedings of International Workshop on Conservation, Characterization, and Utilization of Cocoa Genetic Resources in the 21st Century: Trinidad*, pp. 215–220.

Warren, J. and Emamome, D. (1993) Rodent research in cacao. *Tropical Agriculture* 70(3), 286–288.

Watson, A.G. (1980) The mechanization of coffee production. In: *Proceedings of the 9th International Coffee Research Conference, London, 1980*, pp. 681–686.

Weatherstone, J. (1986) The pioneers 1825–1900. In: Weatherstone, J. (ed.) *The Early British Tea and Coffee Planters and Their Way of Life*. Quiller Press, London, pp. 32–37.

Whitehead, D.L. and Muyila, W.H. (1992) Some new observations on the chemical assessment of quality in clonal teas. *Quarterly Newsletter – Tea Research Foundation of Central Africa* 107, 15–20.

Wight, W. (1962) Tea classification revised. *Current Science* 31, 298–299.

Wight, W. and Gilchrist, R.C.H.H. (1959) Concerning the quality and morphology of tea. In: *Annual Report of Tocklai Experiment Station*. Tea Research Association, Calcutta, pp. 69–86.

Wilkie, A.S. (1994) Irrigation review. Irrigation in context: responses and economic considerations. *Quarterly Newsletter, Tea Research Foundation of Central Africa* 115, 3–6.

Williams, R. (1995) Cocoa integrated pest management. In: *Summaries of the Twelfth International Cocoa Research Conference, Salvador-Bahia, Brazil, 1995*, p. 127.

Willson, K.C. (1965) The design and control of irrigation systems. *Tea* 6(3), 28–29.

Willson, K.C. (1976) Perennial tree crops and well-managed grazing; a partial return to rain forest ecology. In: *Agriculture in the Tropics: Proceedings of the Tenth Waigani Seminar, Papua New Guinea University of Technology, 2–8 May 1976*, pp. 401–407.

Willson, K.C. (1985) Critical levels of nutrients in leaves. In: Clifford, M.N. and Willson, K.C. (eds) *Coffee: Botany, Biochemistry and Production of Beans and Beverage*. Croom-Helm, London, p. 150.

Willson, K.C. (1991) Critical levels of nutrients in the third leaf of a tea shoot. In: Willson, K.C. and Clifford, M.N. (eds) *Tea: Cultivation to Consumption*. Chapman and Hall, London, p. 302.

Willson, K.C. and Choudhury, R. (1969) Fertilisers and tea quality. *Tea* 9(3), 17–19.

Winder, J.A. (1977) Recent research on pollination of cocoa. *Cocoa Growers' Bulletin* 28, 11–19.

Winston, E.C. and Norris, C.P. (1993) Review: development of mechanised coffee production systems in Australia. In: *Proceedings of the 15th International Coffee Research Conference, Montpellier, 1993*, pp. 397–410.

Wood, G.A.R. (1990) The cocoa tree. In: *Exploited Plants*. Institute of Biology, London, pp. 19–24.

Wood, G.A.R. and Lass, R.A. (1985a) Cover crops. In: Wrigley, G. (ed.) *Cocoa*, 4th Edn. Longman, London, pp. 132–134.

Wood, G.A.R. and Lass, R.A. (1985b) Symptoms of deficiencies. In: Wrigley, G. (ed.) *Cocoa*, 4th Edn. Longman, London, between pp. 172 and 173.

Wood, G.A.R. and Lass, R.A. (1985c) Mealybugs and scale insects. In: Wrigley, G. (ed.) *Cocoa*, 4th Edn. Longman, London, pp. 392–397.

Yacob-Edjamo, K., Shimber, T., Yilma, A., Abachebsa, M. and Kufa, T. (1995) Leaf rolling: a drought avoidance mechanism in arabica coffee. In: *16th International Coffee Research Conference, Kyoto, 1995*, Abstract A15.

Yoshioka, M., Aranishi, Y. and Shiraishi, N. (1992) Liquefaction of wood and its application. *FRI Bulletin (New Zealand)* 176, 147–154.

Yow, S.T.K. and Lim, D.H.K. (1994) Green-patch budding on very young cocoa rootstocks and side-grafting of mature trees. *Cocoa Growers' Bulletin* 47, 27–41.

Zhi, L. (1993) Effect of VA mycorrhiza on the growth and mineral nutrition of the tea plant. *Journal of Tea Science* 13(1), 15–20.

INDEX

Note: page numbers in *italics* refer to figures and tables.